ISBN 978-0-282-56105-5
PIBN 10856876

ESTUDIANTE de BAILE AFTER AN OIL PAINTING BY MICHEL JACOBS

A study in overlapping Root Fours with subdivisions in Roots One and Four

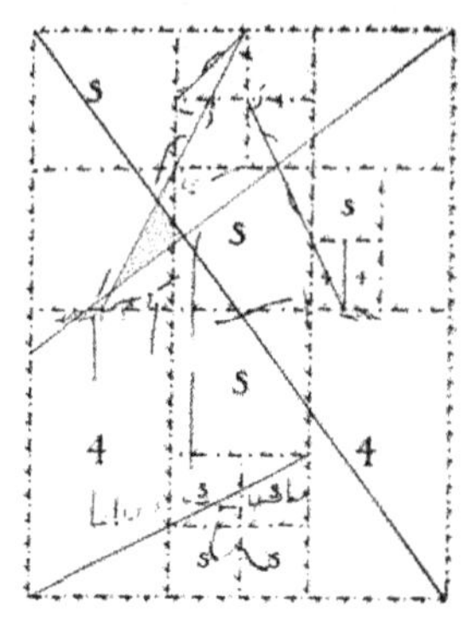

THE ART OF COMPOSITION

A Simple Application of Dynamic Symmetry

BY

MICHEL JACOBS

GARDEN CITY NEW YORK
DOUBLEDAY, PAGE & COMPANY
1926

 PRINTED IN
THE UNITED STATES AT THE COUNTRY LIFE PRESS, GARDEN CITY, N. Y

DEDICATION

Our lives are but a sacrifice; we toil and spin to gain a place in the universe. But if we bequeath to posterity some beautiful thought, some worthy thing, and leave behind the fruits of our labour to help those who follow to make the world more beautiful, we shall have fulfilled our destiny. To those who have gone before and who have left us their life's work, we give our salutations.

M. J.

FOREWORD

HIS book is based on Greek Proportion, which in turn was undoubtedly founded on Nature's own laws. Much of the information was gathered from "Nature's Harmonic Unity," by Samuel Colman, N. A., which was published in 1912, and which was one of the first books published on proportion in nature; from "Dynamic Symmetry: The Greek Vase," "The Parthenon," and other works by Jay Hambidge, published from 1920 to 1924; from "Geometry of Greek Vases," by L. D. Caskey, published in 1922; and from the works of D. R. Hay of Edinburgh, Professor Raymond of Princeton University, and Professor A. H. Church of Oxford.

Many writers have put their own interpretation on this system of composition. This is only natural, when you consider the basic principles from which they have to draw. The variety of compositional layouts are innumerable. I have only attempted to show some of the possibilities, and it is for the artist to work out for himself many more layouts based on this system.

I wish to call to the attention of the reader the fact that this book is only intended as a preliminary study of the great principles of Dynamic Symmetry, and I fervently believe and hope that the readers, after they have perused these pages, will continue their study with the number of books on the subject, and especially the posthumous work by Jay Hambidge called, "The Elements Of Dynamic Symmetry."

Besides the use of this system for artists' composition, I also wish to call to the attention of photographers that Dynamic Symmetry can be used to great advantage, firstly, by using the Transparent Guides described in this book, and secondly, by cutting their photographs so as to conform to dynamic lines and areas, or drawing the dynamic lines on their ground glass.

Advertising agencies and printers will find that their layouts of type matter can be better arranged by the use of this system.

It can also be used by interior decorators, jewellers, and ceramic workers, as well as in other kindred arts.

My thanks are due to my assistants and pupils, Miss Frederica Thomson, Mrs. Eunice Fais, Miss Ruth Radford, and Mr. Louis Amandolare, who have helped me with the illustrations and layouts in this book.

MICHEL JACOBS.

CONTENTS

ILLUSTRATIONS

INTRODUCTION

RCHÆOLOGISTS have long since recognized that the Greeks used Dynamic Symmetry, δυνάμει σύμμετροι in the planning of their temples, statuary, paintings, vases, and other works of art, but artists have been slow to adopt this system mainly because of the belief that it is necessary to understand higher mathematics.

Most books on composition that have been written are books of *"Don't."* I have, therefore, tried to make this book a book of construction rather than tell the things to guard against. I have tried to show how to construct a work of art so as to make the composition a thing of beauty: so that an original conception can be carried out in a harmonious arrangement, as a design or decoration, without which no work of art is worthy of the name.

Another reason that I have taken up this task is to connect Dynamic Symmetry with other forms of composition long since recognized. Undoubtedly, there are many roads: some intertwine; few diverge to such an extent that they cannot be used for the same object. I trust that I have made this book so simple that even a child may be able to master the contents.

On account of the misunderstanding that Dynamic Symmetry is mathematical and difficult to understand, I have taken great pains to leave out any suggestion of an algebraic or geometrical formula. I have even gone so far, in all but the last chapter, to omit letters or numbers to describe lines or angles, for fear that the reader might believe, at first glance, that it was necessary to understand higher mathematics.

One often hears of artists who refuse to be guided by any law or rule of science and who consider that they are a law in themselves. If they were students of psychology, they would see that they are absorbing from others, I might even say copying, perhaps subconsciously, but they themselves would be the first to deny this accusation.

Another peculiar fact, those who do not know the laws of nature and who do not put them into their work often make a great success in their youth through their inherent talent, but in later life fall back in the march of progress on account of their lack of early training and absorbed knowledge.

Painting and drawing have been taught since the days of Ancient Greece by what is known as "feeling." This is all very well, provided that a sound knowledge of construction, of colour, of perspective, and of composition, all based on nature's laws, has been learned and absorbed before "feeling" is permitted to be expressed. Above all, this knowledge, this foundation, must be a part of the artist's subconscious self, so that he does not have to think of rules or methods when he is painting.

If he has not assimilated this knowledge, his work will become stilted, mannered, lacking in charm, spontaneity, and "feeling."

Perspective is used by painters, but they do not necessarily use a ruler, compass, or other forms of measurement. Drawing by eye, they are, however, guided by the rules of perspective. Dynamic Symmetry should be used in exactly the same way.

I have tried to lead the reader from simple compositional arrangement to the more complex, always remembering that the psychological element is to be considered, and that man bases his ideas and feelings of art on nature's laws.

In the last chapter, I have given a few geometrical explanations for those who wish to delve deeper into Dynamic Symmetry. But I must stress that it is not necessary for the artist or photographer to understand the "whys" or "wherefores," except that Dynamic Symmetry is based on nature's laws. If, however, they are intelligent students, they will not be satisfied with working blindly, and will continue the study, in this the most logical and most interesting means of arriving at good compositional forms and layouts.

The little device which I have invented and which I will describe in this book, will help, materially, the artist and photographer to use Dynamic Symmetry in all his work. It will also help the motion-picture director to plan his sets as well as the placing of his principals.

THE ART OF COMPOSITION

١

CHAPTER ONE: COMPOSITION IN GENERAL

OMPOSITION is one of the means to express to others the thought that is in the artist's mind. We can do this with colour, with line, mass, form, or with light and shade—all of which should be combined to bring out more forcibly the idea of the artist.

We must take into consideration at least six things in composing a work of art, whether we paint, photograph, model, engrave or work in any medium by which an idea is expressed in graphic form.

FIG. 1A. COMPOSITION OF MASS

FIG. 1B. COMPOSITION OF VALUE

1. The placement of the different elements.
2. The masses or weights so as to get balance. (See Fig. 1A.)
3. Closely allied to this is the composition of values. (See Fig. 1B.)
4. Composition of line. (See Fig. 1C.)
5. Composition of colour. (See Frontispiece.)
6. Composition of Perspective. This includes ground planning. (See Fig. 1D.)

The placing of a certain thing in a picture or on the stage, which, at first glance, holds our attention, should be the principal object; the eye should then be led to other things which take us from this principal object to

FIG. 1C. COMPOSITION OF LINE

FIG. 1D. COMPOSITION OF PERSPECTIVE

other forms that are associated in a minor key, and which help to express the idea, to be in harmony or act as foils or opposition, and which give to our mind the sense of completeness. Whistler once said, "Nature was made to select from." A work of art is not merely a rendering of nature's planning, but an adaptation by which, in a comparatively small area, one can convey the impression that nature takes the universe to express.

The human mind is a very egotistical thing. Our whole existence is based on our experience with, sometimes, apparently trivial happenings.

Man subconsciously thinks of things which happen to him personally.

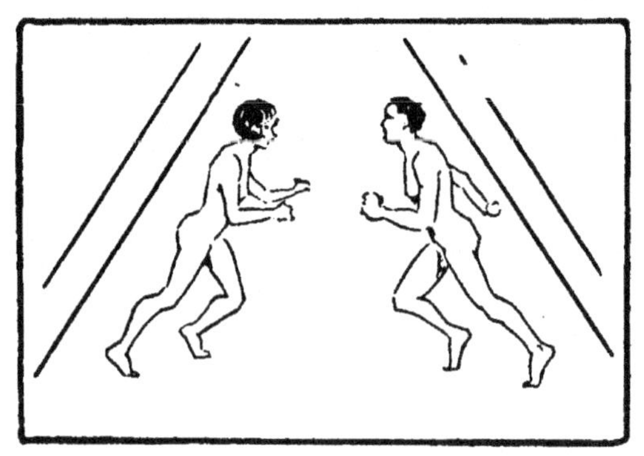

FIG. 2. LINES OF ACTION

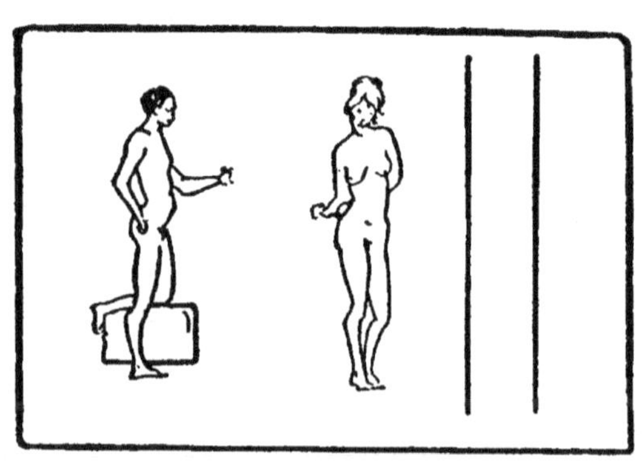

FIG. 3. LINES OF DIGNITY

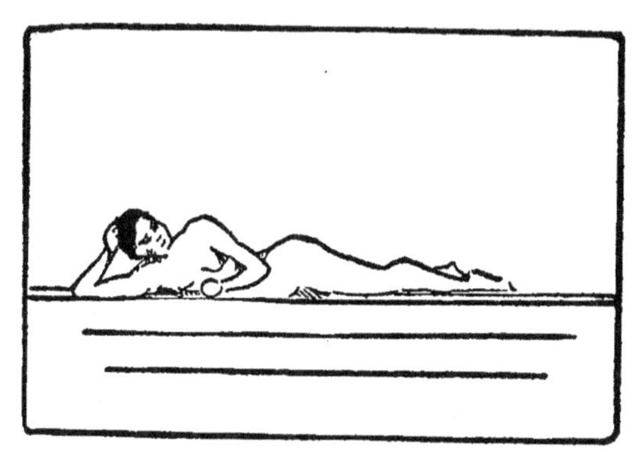

FIG. 4. LINES OF REST

For example, the lines of action are the diagonal (or oblique) lines (Fig. 2); the perpendicular (or upright) lines express dignity and strength (Fig. 3); the horizontal (or lying flat) lines, rest (Fig. 4). This is because man, when running, is in a diagonal (oblique) position, and when perpendicular (upright) he is standing, and when horizontal (lying down) he is asleep or at rest. This is not so when we view the lower order of animals, for the other animals run in an entirely different line of action from what we do, and when they stand, they have also different lines, so we must consider that the lines of composition are based on man's egotistical self. (Figs. 5A, 5B, 5C, 5D.)

FIG. 5A. THE LION

FIG. 5B. ANIMALS IN ACTION

FIG. 5C. THE RAM

FIG. 5D. THE CAMEL

There are certain forms which we unconsciously associate with other ideas. For instance: an arrowhead immediately makes us feel that we should look in the direction in which the point of the arrow is directed. The arrow shaft, however, draws our eye to the feather end rather than to the point, as we like to feel that the arrow is flying away from us rather than toward us. The triangle gives us a feeling of rest and solidity with the idea of pointing upward. The circle gives us the sensation of continued movement. The square gives us the sensation of solidity. The Hogarth lines of grace and beauty give us movement, continuity, and rhythm. The cross gives us the feeling of opposing force, and, on account of its use for religious purposes, the idea will subconsciously be associated with the feeling of piety. As a matter of fact, all form will be associated in the mind of the beholder with previous experiences with that form, and if our composition partakes of these forms, it will express an idea more forcibly than if we did not make use of the subconscious feelings of the onlooker.

In regard to the idea of balance, the seesaw is a very good example. If the board extends equally over each end of the centre rest or fulcrum, it will balance itself. (Fig. 6.) If we put a child on each end of the seesaw, of equal weight, it will also express a perfect balance (Fig. 7), and unless some force or weight is used on one end, it will always stay in this even balance, but if we put a heavier weight on one end and a lighter weight on the other, the heavier end will immediately make the seesaw go down, and likewise raise the other end. (Fig. 8.)

If, again, we put two children on one end and a heavy person (equal to the combined weight of the two children) on the other end, it will also balance evenly. (Fig. 9.)

If we put one of the children toward the centre of the board, the one on the longer end will make the balance go down and the other end go up. (Fig. 10.)

If we lengthen one end of the board and make one end short, and put a child on the short end, and the other end is long enough, it will balance by its own weight. Fig. 11.)

If we shorten the seesaw on the end with the child and lengthen the other end, the child on the short end will be thrown in the air (Fig. 12), and the long end will go down; but if we increase the weight on the short end sufficiently, it will raise the long end to an equal balance. (Fig. 13.)

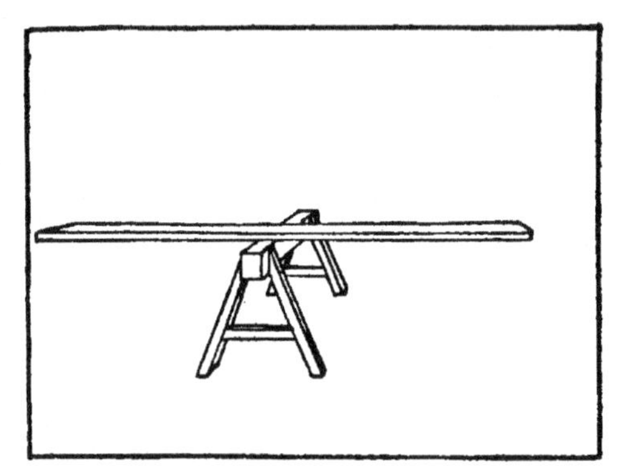

FIG. 6. EVEN BALANCE

FIG. 7. EVEN BALANCE WITH EQUAL WEIGHT

FIG. 8. UNEVEN BALANCE

FIG. 9. EVEN BALANCE WEIGHT SUBDIVIDED

FIG. 10. DISTANT WEIGHT AGAINST WEIGHT NEAR CENTRE

FIG. 11. WEIGHT AGAINST DISTANCE

This gives us a very simple idea of weight, balance, and action, for when the seesaw is on equal balance, we have a feeling of rest, and when one end of the seesaw is down and the other up, we have a feeling that something is going to, or should, happen to make them balance. Each time that we leave the board up or down, we feel that it needs something to complete the action.

We can do this also by placing a stick under the long end to apparently keep it up in the air, and our mind would be satisfied if this support for the long end were heavy enough, in our mind, to hold the board up. (Figs. 14A, 14B.)

FIG. 12. WEIGHT AGAINST DISTANCE UNBALANCED

FIG. 13. WEIGHT SUBDIVIDED AGAINST DISTANCE

All of these examples give you the mechanics of composition, for balance in composition is nothing more or less than a feeling of satisfaction of a completeness of form.

Dynamic Symmetry will help us to arrive at an exact equation of these balances.

If we look at the word DYNAMIC and immediately associate it with the words dynamite and dynamo, we have an idea that it expresses in the word itself, motion. Mr. Hambidge told us that while he named his rediscovery "Dynamic Symmetry," δυνάμει σύμμετροι, the Greeks had themselves already named it by the Greek synonym. It is based on nature's leaf distribution and proportion. Everyone must recognize that nature

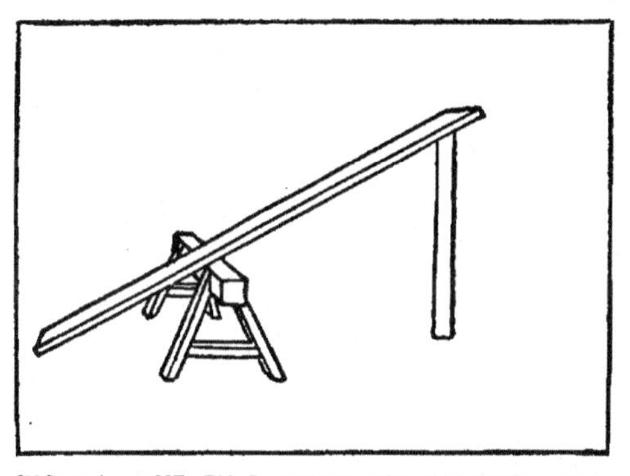

FIG. 14A. UNEVEN BALANCE SUFFICIENTLY SUPPORTED

FIG. 14B. UNEVEN BALANCE NOT SUFFICIENTLY SUPPORTED

does not move, grow, or exist by accident. It is for us poor mortals to discover the secrets of the Master Maker of all things, so as to use them for our own purposes and enjoyment.

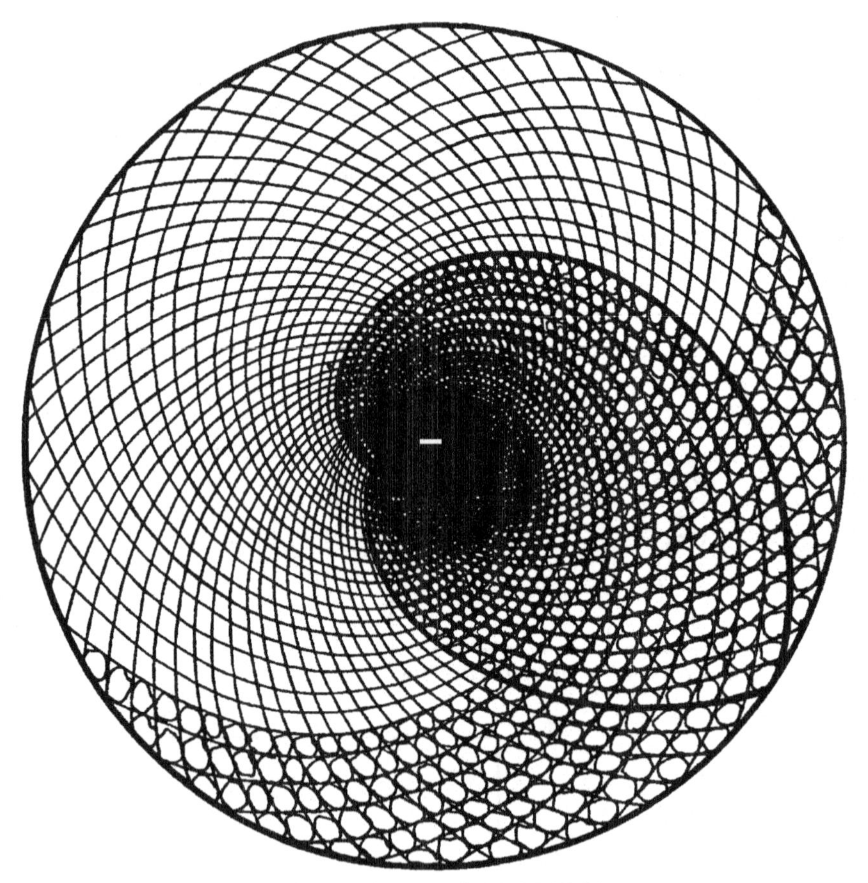

FIG. 15. DIAGRAM OF SUNFLOWER POD

The Royal Botanical Society of London discovered that nature had an order of growth which is based on a peculiar form of numbering, strange, perhaps, to our present civilization.

Taking the seeds of a sunflower pod, which they grew in all sizes, they found these seeds arrayed in a large spiral form, running from the centre, and also in a smaller spiral. (Fig. 15.) Whatever size the sunflower was grown, the seeds of the large spiral numbered in a certain relation to the smaller spiral. They found that if it had fifty-five seeds in the long curve, it always had thirty-four in the short one, and if it had thirty-four in the long curve, it had twenty-one in the short one. It was always in that relation. Then they went further and they found that all nature grew in the same way. That leaf distribution and all vegetable growth was based on this form of numbering, which is called summation.

To explain: if we were to write the numbers 1, 2, 3, 4, 5, 6, 7, 8, 9, 10, 11, etc., we would be adding one each to the number before. If we say 2, 4, 6, 8, 10, 12, we are adding two to each number. If we say 3, 6, 9, 12, we are adding three, and so forth. If we say 1, 2, 3, 5, 8, 13, 21, 34, 55, etc., we are adding the sum of the previous number to the last number enumerated. This is called counting by summation. I shall adapt another word to this means of progression, and call it SEQUENCE. We know the meaning of the sequence of colour. Why shouldn't we say the *sequence of form?*—for that is the meaning of proportion and composition. Using nature as our guide for a means of *sequence of form* will give to us the same feeling of contentment as does sequence of colour.

CHAPTER TWO: DYNAMIC SYMMETRY

YNAMIC SYMMETRY means a certain form of composition—a way of building a picture or other object in good proportion, so that it is pleasing to the eye. Numerous ways of getting composition have been tried since the world began. Dynamic Symmetry is the method by which the Greeks built their temples and their gods. In the Middle Ages, a different form of composition was used. The Japanese, Chinese, and others used different forms. Remember, while Dynamic Symmetry is a wonderful thing, it is not the only way of getting a good composition. Dynamic Symmetry really means a composition of spaces or areas, one in harmony or sequence with another. There is a composition of line, of space (notan, as the Japanese call it), as described by Dow, and of mechanical balance as described by Poor. An artist who wishes to express action, animation, or movement, will find that Dynamic Symmetry answers better for all his requirements.

This form of composition is a composition of action, which does not necessarily mean that a figure has to be in motion, but simply that the lines or masses express motion. In Dynamic Symmetry the compositional forms express motion, as in Figs. 1 to 15. Opposed to this form of composition is one called static, or still—a bi-symmetrical composition is often a static composition.

Dynamic Symmetry is really not difficult to learn providing you look at it in a simple, common-sense way. Remember, it is not one man's theory of composition—it is the Greek form of composition. A Grecian would have said, for example, this page was composed in Root two—as we say so many inches high and wide. Root One was a square, and from this they constructed Roots Two, Three, Four, and Five, etc. Ours is lineal measure and theirs is a measure of space.

Dynamic Symmetry composition is not a thing that will make you mechanical, as it bears the same relationship as perspective to composition. If you know the laws of perspective, you draw the perspective free hand.

By drawing a square, you make Root One. The diagonal of Root One is the length of Root Two; the diagonal of Root Two is the length of Root Three; the diagonal of Root Three is the length of Root Four; and the diagonal of Root Four is the length of Root Five, etc. If the Greeks wanted to measure the ground of a temple, they would say it was so many Root One's, Two's, Three's, or other roots. If you paint a picture and use one of the roots for your size of canvas, you will have a well-proportioned form to start with, so far as proportion of space to be covered.

Inside of this form we may wish to place a composition, and we want to know where the objects are to be placed. One should think of composition as a means of expressing an idea based on a psychological reaction of the onlooker, and this reaction is based on a previous experience, either physical or mental.

If we were to draw two circles in very close proximity to each other, almost touching, our minds would immediately feel that two revolving bodies were about to collide, as in Fig. 16. This is what is known as tangents in all forms of composition. This same sensation is given when you look at a circle or other form about to strike the picture frame, as is illustrated in Fig. 17.

If we overlap the circle or cut it with the frame, our mind is immediately associated with an idea that the two which overlap have passed, one over the other (Fig. 18), and that the one cut by the

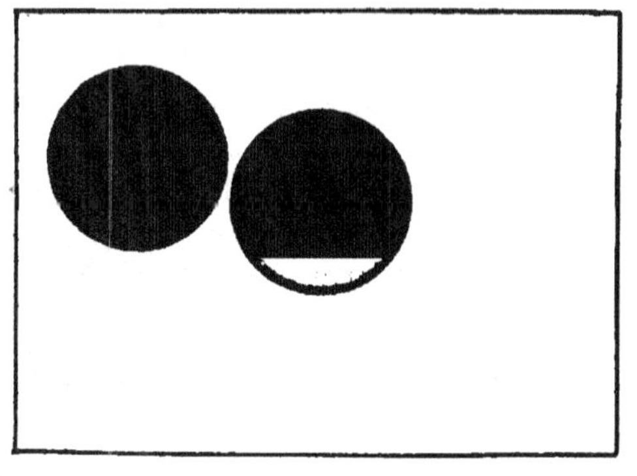

FIG. 16. TANGENTS ABOUT TO COLLIDE

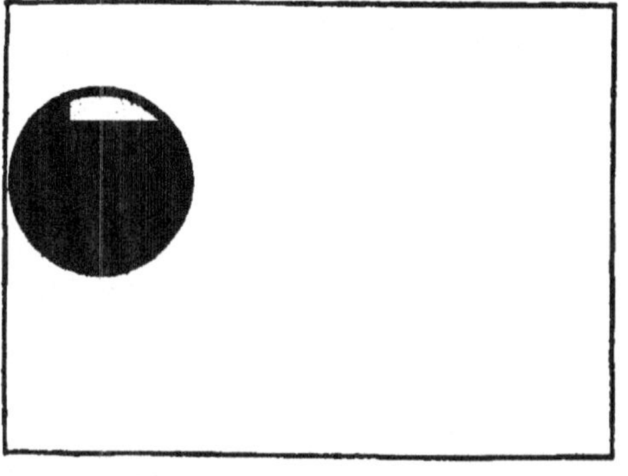

FIG. 17. TANGENT NEAR EDGE

FIG. 18. TANGENTS OVERLAPPING

frame continues indefinitely and suggests the infinite, which is always interesting to the beholder. (Fig. 19.) In other words, onlookers like to believe subconsciously that they are completing the picture themselves. It is another form of psychology. The old saying that "things must be taught as things forgot" should be carried out in the pictorial arts so that the beholders believe they "had a hand" in the completion of the work. This is all, of course, in the subconscious mind. The painter must always figure on the psychological effect of his arrangement.

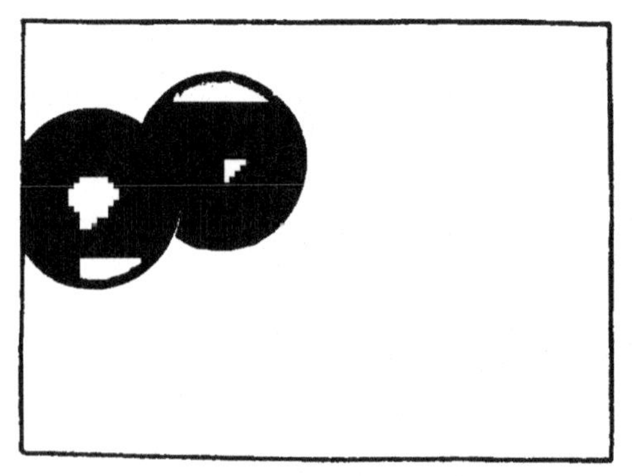

FIG. 19. TANGENTS OVERLAPPING AND CUTTING EDGE

A very good example of the psychological reaction of the beholder is the seesaw. If one end of the seesaw is portrayed in the air, with the support in the centre, our mind immediately waits for it to come down in the return action. (Fig. 8.) Especially is this so if the end in the air is more heavily weighted than the one on the ground. Whereas, if the seesaw is more heavily weighted on the end resting on the ground, and the part that is in the air is lighter or without weight, we do not wait for the return of the action, we are contented with a sense of finality, and that there will be no chance of it coming down again without some added force or weight. This should also be carried out, as was explained before, in your picture. One is almost tempted to say that composition is nothing more or less than the psychological reaction of the beholder; that they are satisfied or dissatisfied with the action that has taken place or been expressed.

On the other hand, one might wish purposely to express an unpleasant theme and would wish to use the tangent, perhaps, to give an idea of a catastrophe. (Fig. 20.) Take, for example, two fighters. If a fist of one were about to strike the other's head, it would form a tangent and would be the correct form for that action. (Figs. 21A, 21B.)

As I have said, all nature grows in a certain relation, and this order of growth in either vegetable or animal life has been found to be in the relation in whole numbers of about 1, 2, 3, 5, 8, 13, 21, 34, 55, etc. This is not so exact, however, as the relation of the numbers 118, 191, 309, 500, 809, 1309, 2118, 3427, etc. We will use for the present, though a little inaccurate, the smaller approximate whole number summation of 1, 2, 3, 5, 8, 13, 21, 34, 55, etc., to show the relationship of nature to composition. If we draw an oblong which measures 5 inches by 8 inches, and then draw a diagonal, or hypotenuse, from the two far corners, and crossing this diagonal with a line (one end of which rests in the corner, and which crosses the diagonal line at right angles) continuing through to the opposite side of the oblong, we shall have drawn an oblique cross in the oblong.

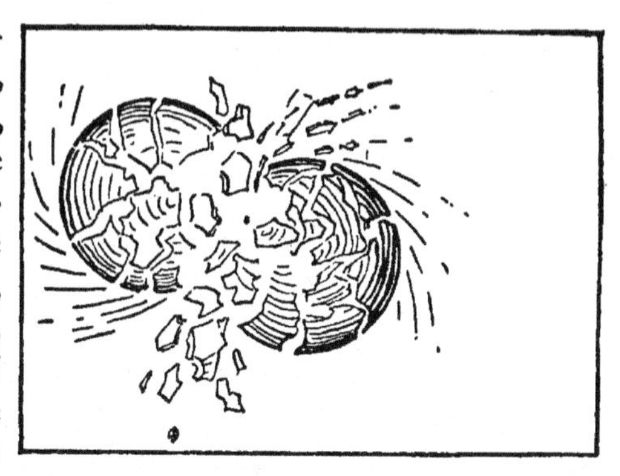

FIG. 20. THE SUGGESTED EFFECT OF TANGENTS

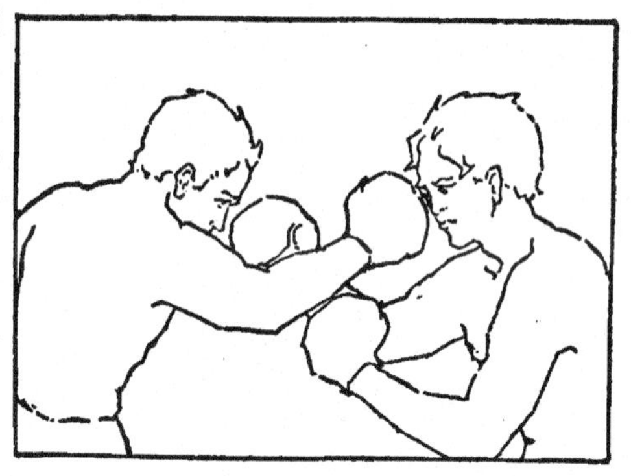

FIG. 21A. TANGENTS THAT ARE NECESSARY

By drawing a line parallel with the side, where the short crossing line touches the sides of the oblong, so as to form a square on one end, you will have produced the original form of the oblong, but in a smaller proportion or sequence on the other end. As the original form measured 5 by 8, the smaller form will measure 3 by 5, and if we draw another line across where the diagonal meets, we shall have a smaller form which will measure 2 by

FIG. 21B. TANGENTS THAT ARE NECESSARY

THE INDIAN

AFTER AN OIL PAINTING BY MICHEL JACOBS

In Root Four

THE DUCHESS AFTER AN OIL PAINTING BY MICHEL JACOBS

In Root Two

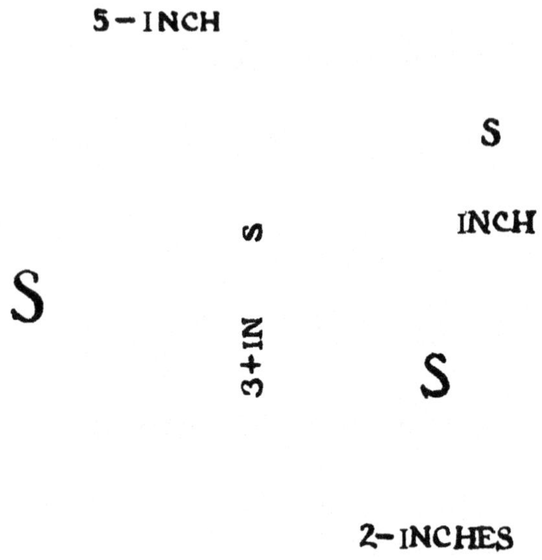

5-INCH

S

INCH

S

S

3+IN

S

2-INCHES

FIG. 22. SEQUENCE OF AREA IN WHOLE NUMBERS

3, and if we draw again another line, we shall have a smaller form which will measure 1 by 2 inches. (Fig. 22.) By this method, you will see that you have made smaller forms in the large rectangle, or forms in sequence which will measure in the summation of 1, 2, 3 5, 8; the same as if we strike high "C" or low "C" on the piano. It is the same sound, only one is, so to speak, greater in its vibrations than the other.

CHAPTER THREE: DIFFERENT ROOTS OR FORMS

PROPORTION OF PICTURES

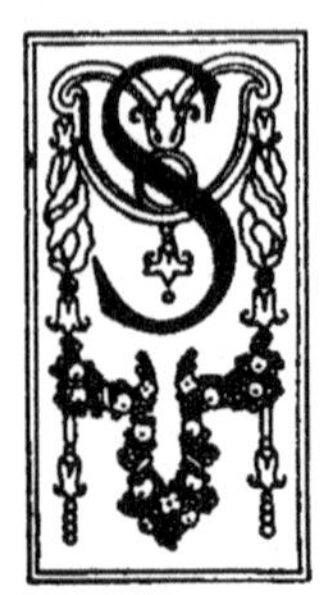

STARTING at the beginning to study Dynamic composition, one must first learn the so-called roots, these are nothing more than squares and oblongs of different proportion. While one can make any number of roots, I have only shown up to Root 26 in this book, and have used only six roots to explain this form of compositional proportion, namely, Roots One, Two, Three, Four, Five, and the Whirling Square Root.

To simplify the descriptions of roots and crossing lines, we shall designate hereafter:

The Diagonal: A line drawn from opposite corners or any oblique line. (Fig. 23.)

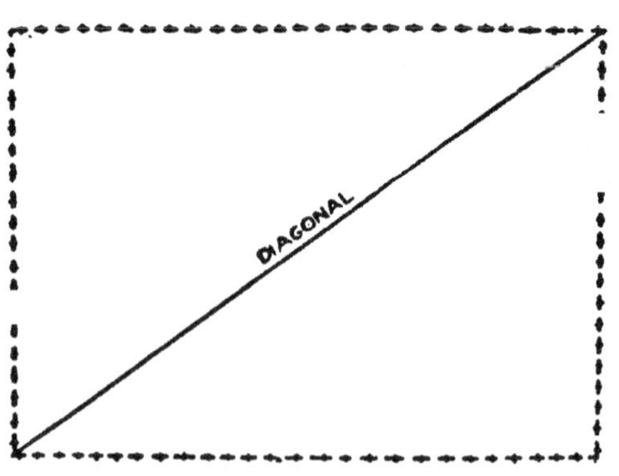

FIG. 23. ROOT TWO SHOWING THE DIAGONAL

A Crossing Line: A line drawn from one corner to the outside edge, crossing the diagonal (Fig. 24) at right angles.

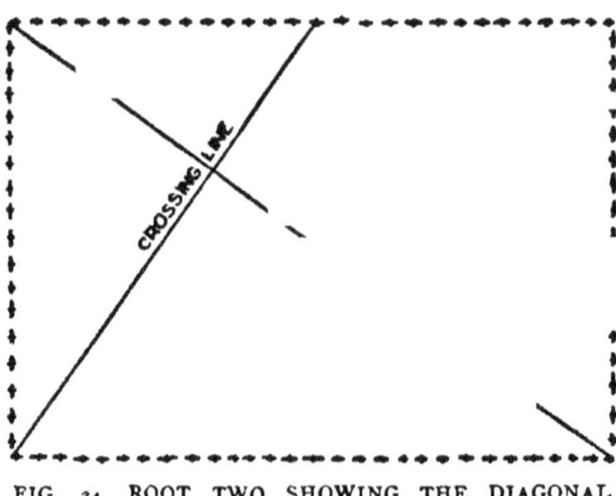

FIG. 24. ROOT TWO SHOWING THE DIAGONAL AND CROSSING LINE

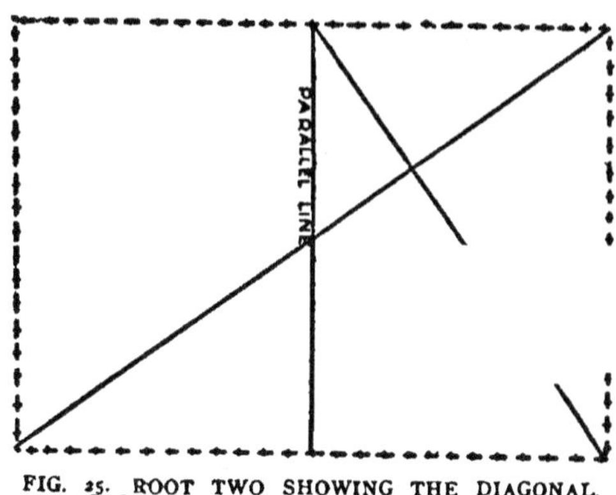

FIG. 25. ROOT TWO SHOWING THE DIAGONAL, CROSSING LINE AND PARALLEL LINE

Parallel Lines: Lines drawn straight across the form, parallel to the sides or ends, at the point where the crossing line touches thé outside edge through any intersection (Fig. 25) or any line parallel to any root boundary. (Fig. 26.)

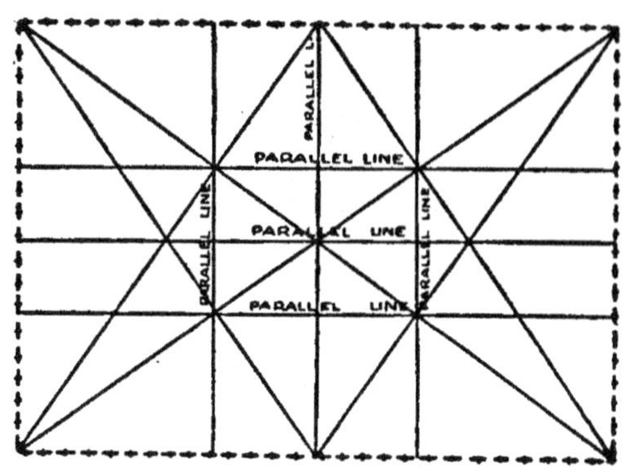

FIG. 26. ROOT TWO SHOWING THE PARALLEL LINE IN ALL DIRECTIONS

To avoid confusion and to enable the reader to pick out the roots at sight, in the different layouts I have adopted the following symbols to designate each root.

The Root One will always be designated by a line made with a series of dots: . . . .

The Root Two will always be designated by a line made with a series of crosses: + + + + + + +

The Root Three will always be designated by a line made with a series of dashes: — — — — — — —

The Root Four will always be designated by a line made with a series of angles:

The Root Five will always be designated by a line made with a series of wavy dashes:

The Whirling Square Root will always be designated by a line made with a series of spirals:

The diagonal and crossing lines as well as lines which do not form another root will be designated by a straight continuous line.

By this means, we shall designate and the reader will be able to distinguish readily the roots which are contained in the grand mass.

If there are two roots, one overlapping the other, for example, a Root One containing two Root Four's, they will be designated by a line made with a series of dots interposed with the angle:

or, as another example, if a Root One contained a Whirling Square Root it would be designated by a line made with a series of alternate dots and spirals:

Another example would be, if a Root Three contained two other roots, a Root Two and a Whirling Square Root, it would be designated by a cross, a spiral, and a dash: + — + — +

(See Fig. 27 for all of these symbols.)

In addition to these markings, wherever possible, I have also designated the root by the number of the root. The Root One I have marked S, which stands for a square, and the Whirling Square will be known as W.S.; the other roots by the number of the root and also in the captions under each illustration the roots used.

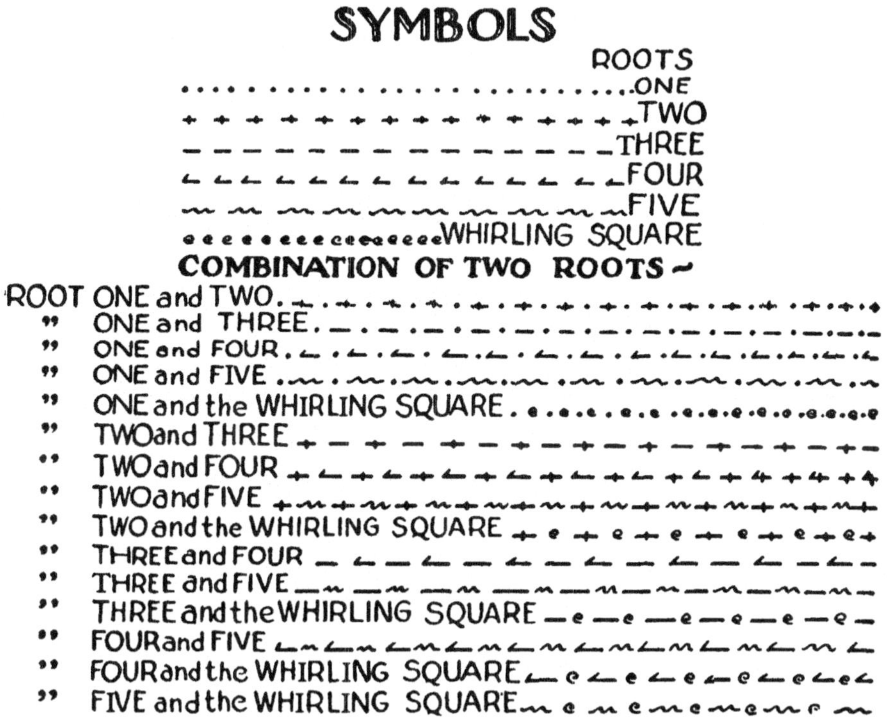

SYMBOLS

ROOTS

ONE

TWO

THREE

FOUR

FIVE

WHIRLING SQUARE

COMBINATION OF TWO ROOTS

ROOT ONE and TWO

" ONE and THREE

" ONE and FOUR

" ONE and FIVE

" ONE and the WHIRLING SQUARE

" TWO and THREE

" TWO and FOUR

" TWO and FIVE

" TWO and the WHIRLING SQUARE

" THREE and FOUR

" THREE and FIVE

" THREE and the WHIRLING SQUARE

" FOUR and FIVE

" FOUR and the WHIRLING SQUARE

" FIVE and the WHIRLING SQUARE

FIG. 27. SYMBOLS OF THE DIFFERENT ROOTS

Root One—It is easily understood that if you multiply one side of a unit square by the other side, you will get the unit one: $1 \times 1 = 1$; so we say the square is Root One (as in Fig. 28.) (See Glossary for definition of Square Root.)

FIG. 28. ROOT ONE

Root Two—If we measure the diagonal, or hypotenuse, of a square we get the *length* of Root Two, and the side of the square itself is the other side; for example, if we take a square measuring 3 inches (7.65 centimetres), we find the diagonal measures about $4\frac{1}{4}$ inches (10.20 centimetres). By making an oblong measuring about $3 \times 4\frac{1}{4}$ inches (7.65×10.20 centimetres), we have constructed a Root Two rec-

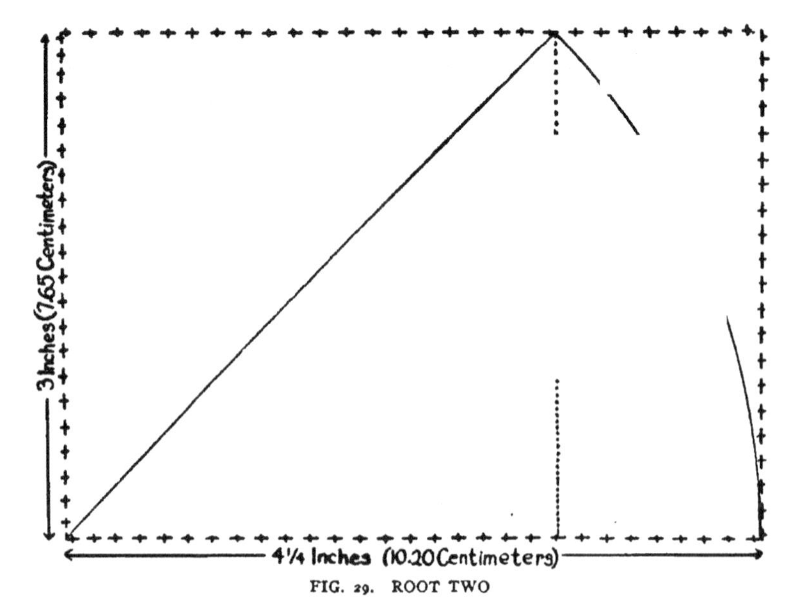

FIG. 29. ROOT TWO

tangle, or oblong; if the measurement is done with a compass, you do not need to know the number of inches. By simply putting one point of the compass on the corner of the square and the other point on the opposite far corner, you will have the length of Root Two. (Fig. 29.)

Root Three—By putting the compass points on the two opposite far corners of Root Two, you will find out the length of Root Three, and of course, the width will be the width of the same square. This Root Three will measure about 3 × 5¼ inches (7.65 × 13.40 centimetres) as in Fig. 30) if we use the same base line of three inches.

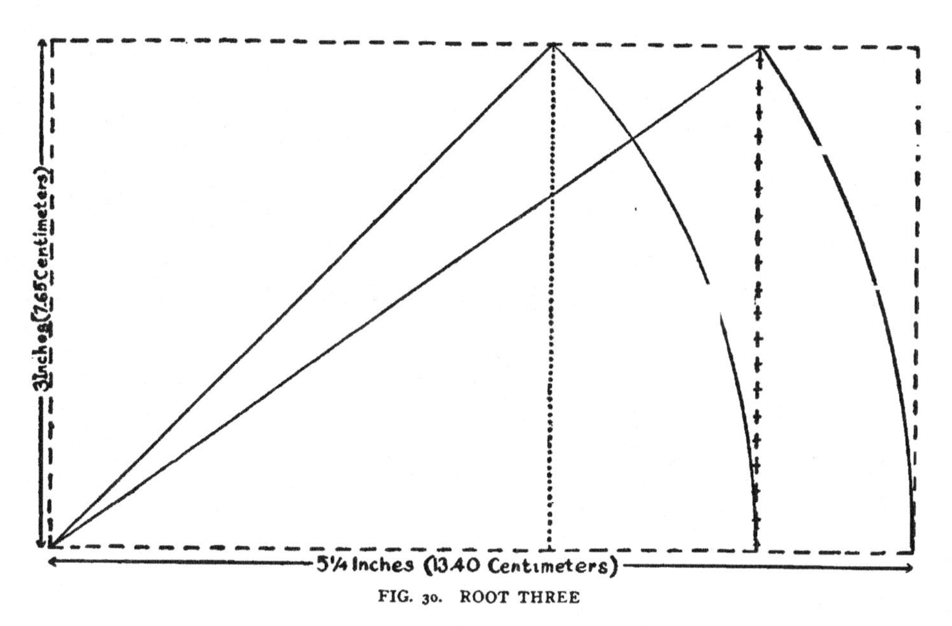

FIG. 30. ROOT THREE

Root Four—By again putting the compass points on the two opposite far corners of Root Three, you will find out the length of Root Four, and again the width will be the size of the original square. This Root Four will measure about 3 × 6 inches (7.65 × 15.30 centimetres), using the same square. Two Root One's equal a Root Four (as in Fig. 31), and one half of a square measures a Root Four.

Root Five—By once again putting the compass point on the two opposite far corners of Root Four, you will find out the length of Root Five, again using the width of the original square: 3 inches. This Root Five will measure about $3 \times 6^{11}/_{16}$ inches (7.65×17.70 centimetres) as in Fig. 32. All the other roots can be drawn in the same manner.

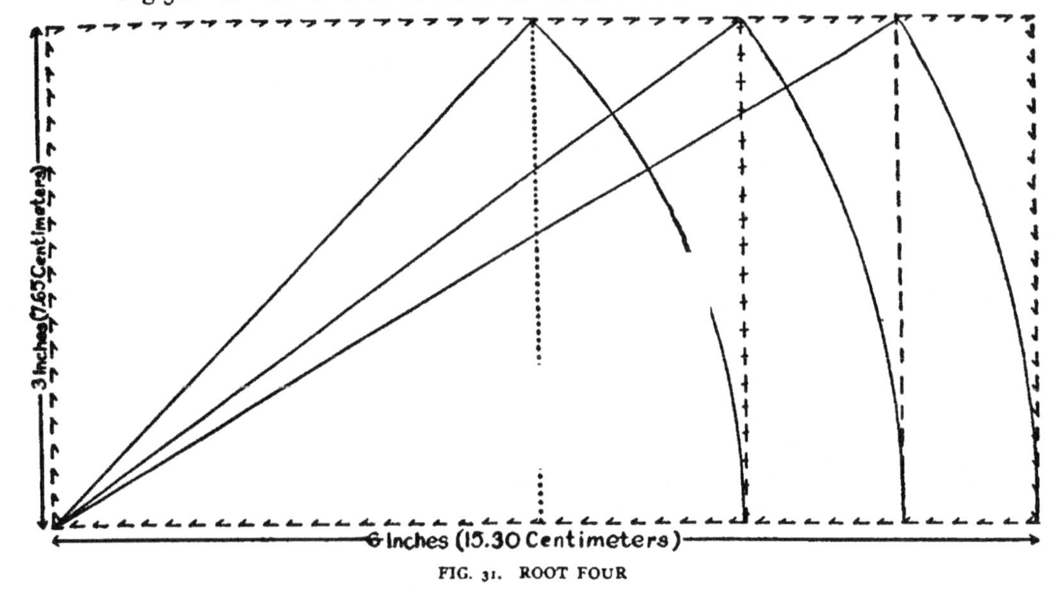

FIG. 31. ROOT FOUR

6 11/16 Inches (17.70 Centimeters)

FIG. 32. ROOT FIVE

THE WHIRLING SQUARE ROOT—This root is a little different from all of the foregoing roots. We find out this by taking a square and marking off half of one side; we measure the diagonal of this half, using the same 3-inch square. This diagonal will measure 3⅜ inches (8.65 centimetres). By adding the half of the square to the length of this diagonal, you will have the length of the Whirling Square Root; using again the width of the original square for the small end, the Whirling Square Root will measure 3 × 4⅞ inches (7.65 × 12.50 centimetres), as in Fig. 33.

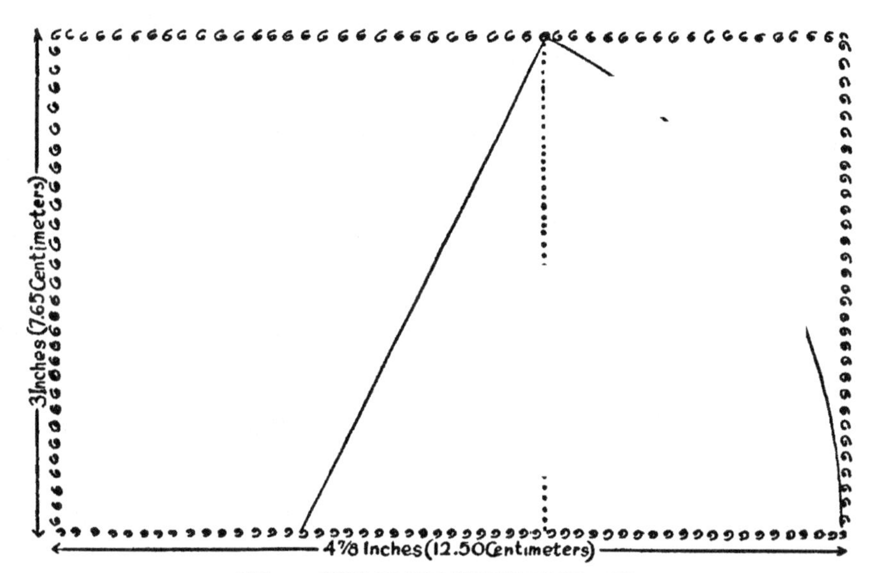

FIG. 33. ROOT OF THE WHIRLING SQUARE

Another way to form the root of the Whirling Square is to take a square, and from the centre of one side draw a half circle; this arc will touch the corners of the square. This will give you a Whirling Square Root on each end, and the whole form will be a Root Five. By taking away one of the small Whirling Square Roots or oblongs, you will have a Whirling Square Root, which is a square and a Whirling Square. So, you see, the Root Five and the Whirling Square Root are closely related, as in Fig. 34.

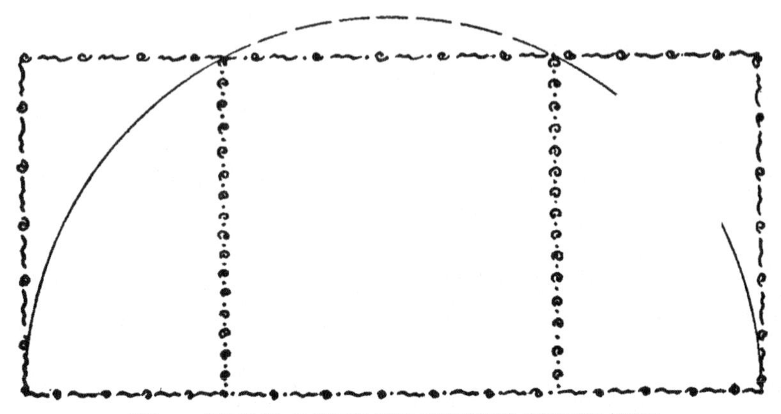

FIG. 34. ROOT FIVE WITH THE WHIRLING SQUARE ROOT

Putting the Roots Outside of a Square—All of the roots can be placed outside of a square by drawing a square, taking a diagonal of the square or Root One and laying it along the base of the square and drawing the oblong or rectangle; again laying down the diagonal of Root Two, you will get the length of Root Three; again laying down the diagonal of Root Three you will have the length of Root Four; and again laying down the diagonal of Root Four, you will have the length of Root Five, as in Fig. 35.

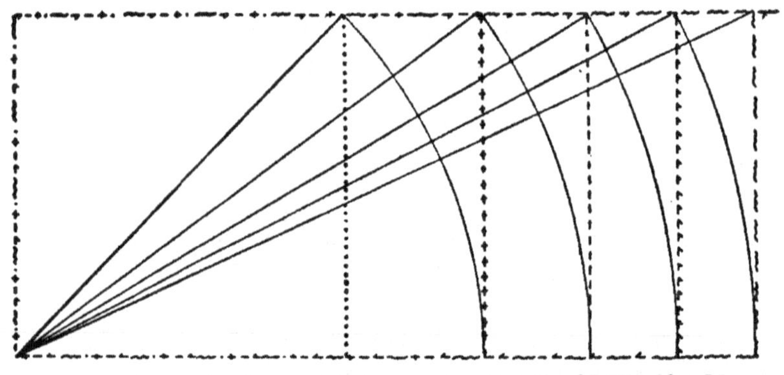

FIG. 35. ROOTS TWO, THREE, FOUR, AND FIVE OUTSIDE OF THE SQUARE

PUTTING THE ROOTS WITHIN A SQUARE—All of the roots can be placed within a square by the following method. Draw a square and with a compass make a quarter circle, called an arc, the two ends resting in opposite corners, and by drawing a diagonal from the opposite two corners, you will cross this quarter circle or arc at the centre; then, by drawing a line parallel with the top side and base of the square, you will have formed Root Two within the lower part of the square, as in Fig. 36

By again drawing a diagonal line from the corner of the square to the corner of the Root Two oblong thus formed within the square, you will draw the parallel line to form Root Three where this diagonal crosses the quarter circle. Again drawing the diagonal line from the corner of the square to the corner of this Root Three, you will know where to draw the parallel line to form Root Four. And again drawing the diagonal line from the corner of the square to the corner of the Root Four rectangle thus formed, you will have formed Root Five. Wherever the diagonal crosses the arc is the place to draw the parallel line to form the root as in Fig. 36.

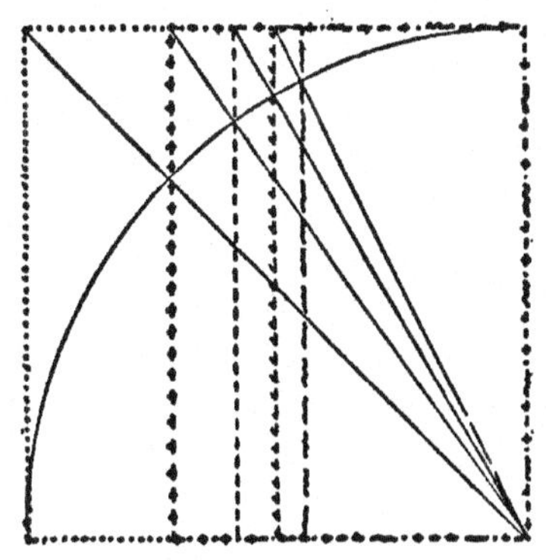

FIG. 36. ROOTS TWO, THREE, FOUR, AND FIVE INSIDE A SQUARE

Any of these root rectangles, or oblongs, will be found to be a good proportion for your canvas, or board, on which to draw or paint your pictures, and as a guide for the photographer in cutting his print.

It must be remembered that Roots One, Two, Three, Four, Five, etc., bear a relationship one to the other. For example, Root One, we have seen, contains all the other roots, and, likewise, Root Five contains all the other roots, so that, by using two or more of these roots, one within the other, we have a perfect sequence of area. In the following chapters I will show how this relationship of the different roots is used to form various compositional layouts.

A very simple method of finding out what is the root of any shape would be to use a compass on the small end, measure off the square on the long side, point off Root Two, then, with a compass, take the diagonal of Root Two, lay off Root Three, etc.

If you wish to measure with a ruler, you could take the measurement of the long side of the oblong or rectangle and divide it by the short side.

This would give you the symbol number of the root, or as it is called the Reciprocal. For example, if the oblong measures 3 by 4¼ and you divided this by 3 (the length of the short side), it would give you 1.41 +. You would know this to be a Root Two rectangle. (See Proportion of Roots, Chapter Sixteen.) Another example: if the rectangle measured 5 by 7.071 inches and you divided it by the short side, namely 5 inches, the result would again be 1.4142. This would show that it was also Root Two. If you have a form which measures 3 by 6 inches and you divide the long side by the short side (6 divided by 3 is 2.000), you know that this is the number which shows that the proportion is a Root Four.

Inversely, if you wish to know the measurement of the long side of any root and you know the dimension of the short side, multiply the short side by the number of the root that you wish to use, called the Reciprocal, as shown in Chapter Sixteen; this will give you the length of the long side. These measurements are only approximate, and are used only to make the idea more comprehensible to the beginner.

CHAPTER FOUR: POINTS OF INTEREST

SIMPLIFIED FORMS AND LAYOUTS IN ROOTS ONE, TWO, THREE, FOUR, AND FIVE

IN THE preceding chapter, we have found out the proportion to make our picture forms that are pleasing to the eye in the major shape. We now shall find where to place the principal points of interest.

By taking Root Two instead of starting with Root One, we shall have a simpler form to explain the beginning of Dynamic Symmetry composition. As we explained in the previous chapter, Root One is a combination of two Root Four's and can be divided up into many more forms, and is a little harder to study at the outset than the Roots Two, Three, and Four. So, as I have said above, we will start with finding out the principal points of interest in Root Two.

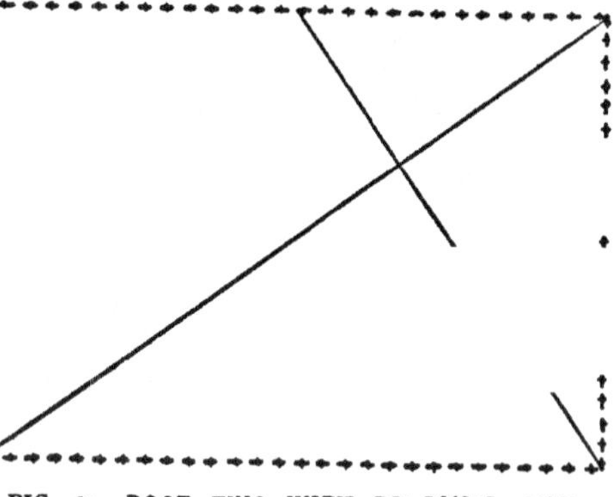

FIG. 37. ROOT TWO WITH DIAGONAL AND CROSSING LINE

By drawing an oblong which we know to be Root Two, as is explained in the previous chapter, we take the diagonal of a square; this diagonal would be the length of Root Two, as is illustrated in Fig. 29. After drawing a diagonal from the two far corners, cross it with an oblique line one end of which rests in the corner, crossing the diagonal line at right angles, continuing through to the opposite side of the oblong, as is illustrated in Fig. 37: where these two lines cross will be one

FIG. 38. THE OLD WOMAN WHO LIVED IN A SHOE

of the artistic centres of the rectangle. It will be, of course, understood that the diagonal may be drawn from the opposite corners, and the crossing line, or, as it is known, the line "squaring the diagonal," may be drawn from any one of the four corners. Also, the oblong may be upright or lying on its side, as is illustrated in Fig. 38, Fig. 39, and Fig. 40.

FIG. 39. COUNTRY ROAD

Any one of these points of interest may be considered as principal points of interest. If we were to draw the diagonal and cross it on the top and bottom, we would have two points of interest, one of which we will call the principal and the other the secondary point of interest in SEQUENCE. (Fig. 41.) Again, we can take the oblong and draw on it both diagonals and all four of the short crossing lines to make it look like Fig. 42.

You will notice on this illustration that I have designated one point the principal point of interest, another the second point of interest, another the third point of interest, and another the fourth point of interest. To explain this matter further, I would refer you to Fig. 43 which shows you a simple composition based on the idea of taking the points of interest in SEQUENCE.

We have shown now a simple way of getting points of interest in the flat plane in Root Two. This same layout may be applied to Roots Three, Four, and Five (as in Figs. 44, 45, 46).

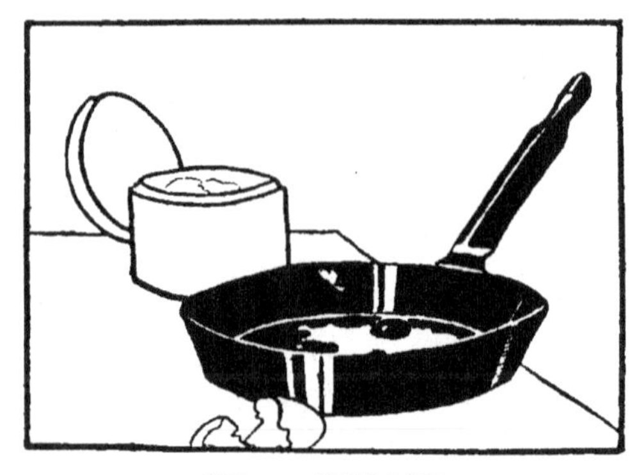

FIG. 40. STILL LIFE

When placing the object in the principal point of interest, it is very good to follow, more or less, the lines of the compositional construction, namely,

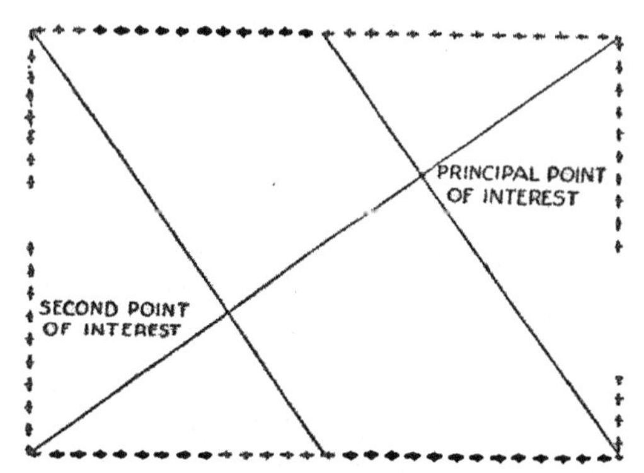

FIG. 41. PRINCIPAL AND SECOND POINTS OF INTEREST

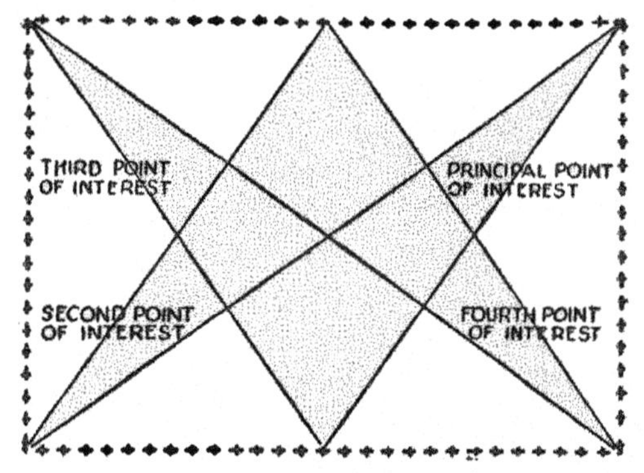

FIG. 42. PRINCIPAL, SECOND, THIRD, AND FOURTH POINTS OF INTEREST

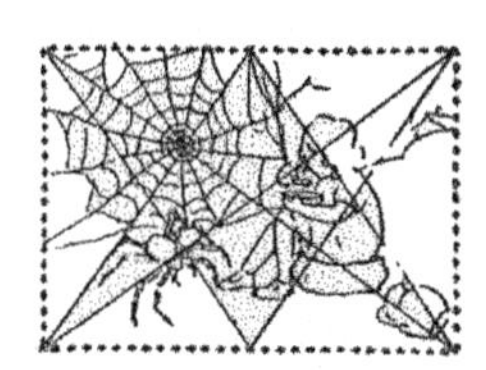

FIG. 43. LITTLE MISS MUFFET

THE CRYSTAL GAZER AFTER AN OIL PAINTING BY MICHEL JACOBS

In Root Two

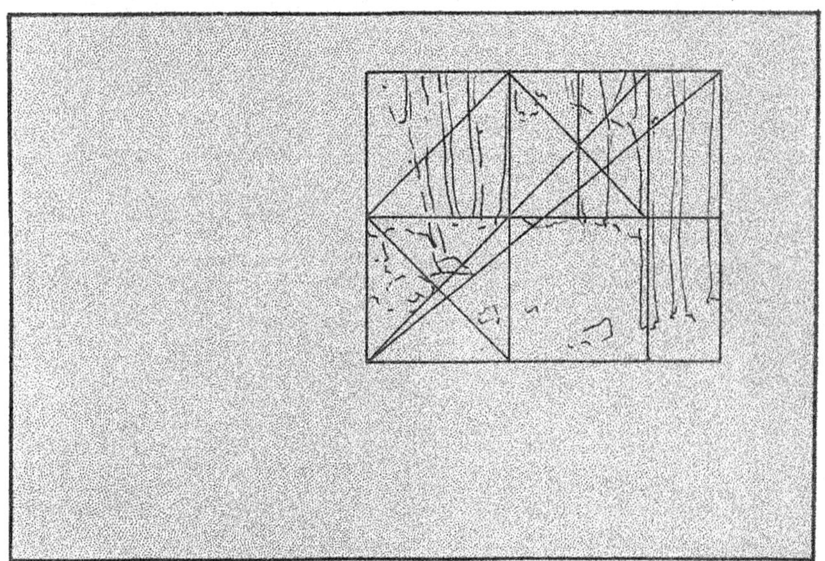

THE HILLSIDE AFTER AN OIL PAINTING BY MICHEL JACOBS

A form less than Root Two

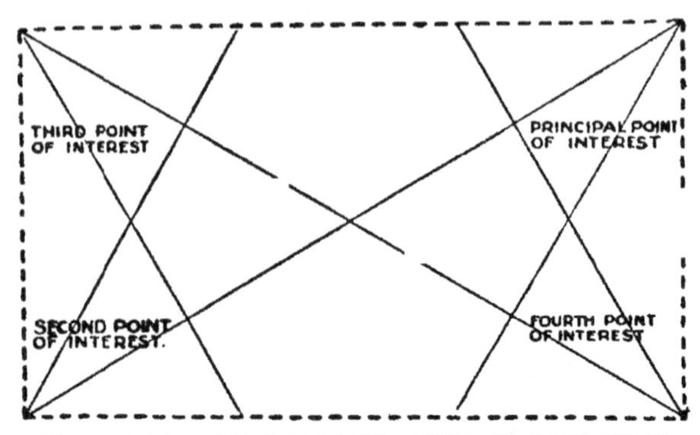

FIG. 44. ROOT THREE SHOWING POINTS OF INTEREST IN SEQUENCE

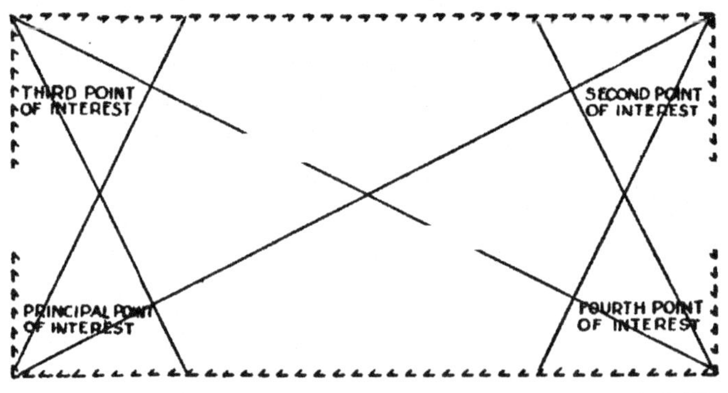

FIG. 45. ROOT FOUR SHOWING POINTS OF INTEREST IN SEQUENCE

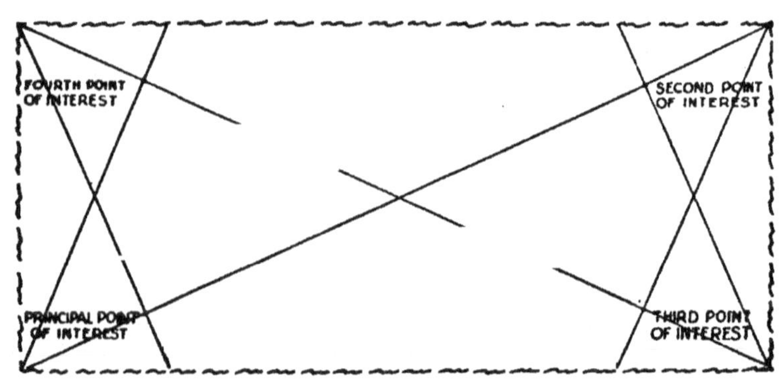

FIG. 46. ROOT FIVE SHOWING POINTS OF INTEREST IN SEQUENCE

the diagonal and the short crossing line. It is not necessary, however, to make the composition exactly on these lines. The composition would become too mechanical if this were done. Perspective is used by an architect in a different way than by a painter. The architect's drawing is hard, cold and mechanically perfect, whereas the painter's perspective is free and bold and more artistic. Remember that Dynamic Symmetry has the same relation to composition and proportion that perspective has to any painting, photograph, or stage set.

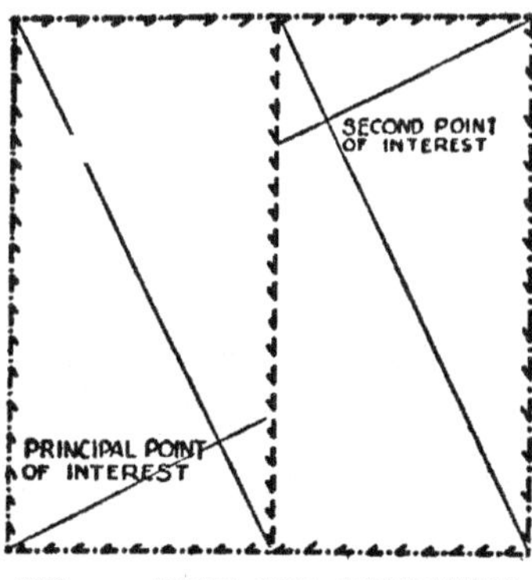

FIG. 47. ROOT ONE CONTAINING TWO ROOT FOUR'S

The Root One, as I explained before, contains two Root Four's. This could be used in a number of ways; the more complicated ones I will take up later. Dividing the square in half, and drawing the diagonal through each of the halves, and squaring the diagonal, as is illustrated in Figs. 47 and 48, would give you two points of interest, which could be considered as principal and secondary. I have also shown a simple sketch based on this idea.

Continuing to make this Root One useful for a composition with the two Root Four's which it contains, divide each one of these Root Four's, the same as we did with the Root Two, with both the diagonals of each Root Four and the crossing lines; we now have eight places to select from as our principal points of interest and seven points of interest in SEQUENCE. We also have a number of crossing lines which we can use more or less in our composition, as is illustrated in Figs. 49 and 50.

FIG. 48 THE BATHER

Another way that we could use Root One would be to draw the diagonal line, and then, with a compass, draw a quarter circle to find out the smaller roots inside of a square, as we explained in Chapter Three, we would have lines as illustrated in Fig. 51.

Now, if we draw a line where the quarter circle crosses the diagonal, as if we were going to find out the Root Two (as is explained in Fig. 36, Chapter Three), and if we do this on both sides and ends, we shall have a layout of lines for a composition which will be as is shown in Fig. 52. We can

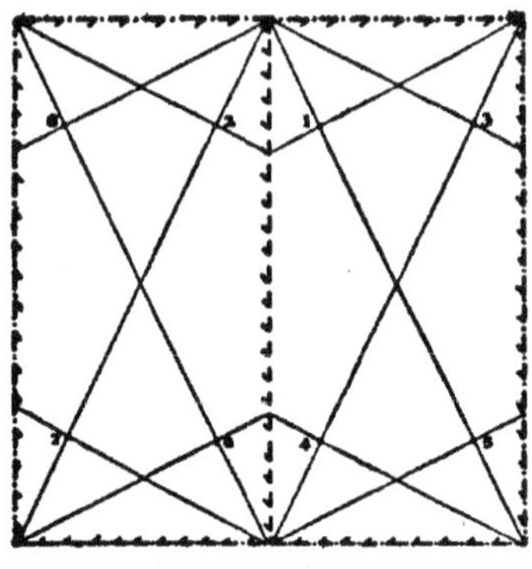

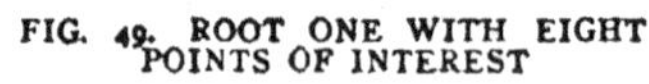

FIG. 49. ROOT ONE WITH EIGHT POINTS OF INTEREST

FIG. 50. THE DANCER

select any one of the points as a principal point of interest, and the other ones to be points of interest in SEQUENCE, as is illustrated in Fig. 53. There are many different ways of using all the roots, which will be explained later. The few compositional layouts which I have given are for the beginner.

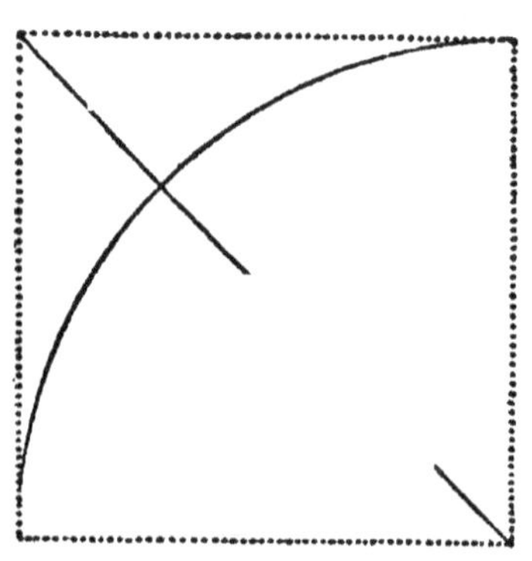

FIG. 51. ROOT ONE WITH A QUADRANT ARC

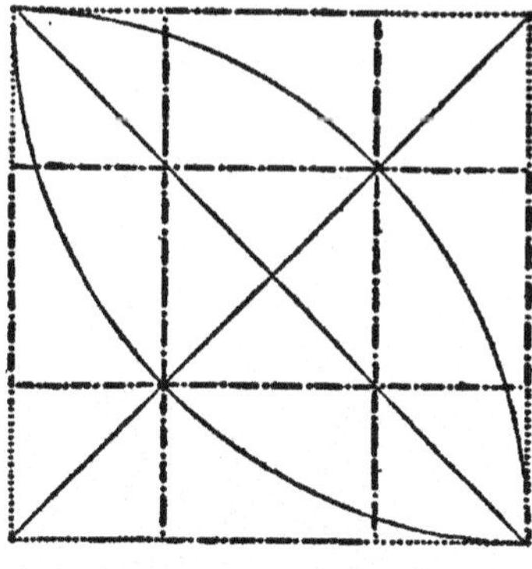

FIG. 52. ROOT ONE WITH TWO QUADRANT ARCS AND PARALLEL LINES

FIG. 53. THE OLD FASHIONED GARDEN

I wish here to suggest a method whereby the different roots and layouts can be kept for future reference, which will help the painter to correct the original drawing without again making the layout for the original picture.

If you take a number of small transparent guides about two or three inches in width, each one with the different roots drawn upon it with waterproof drawing ink, and draw the crossing lines, the diagonal, and parallel lines which you wish to use, now holding this within a few inches of the eye, and standing off a few feet from the picture, you will be able to judge the corrections to be made in your painting to make it conform nearer to the root line or mass. These small guides can be kept for future reference for other compositions. This will do away with the task of each time drawing a separate layout on the canvas. In Fig. 54 is an illustration showing how this system can be used.

For those who desire them the Pri-matic Art Company, of New York, have made up, under my direction, all the layouts contained in this book printed on transparent guides.

ANNA PAVLOWA AND MICHAEL MORDKIN

AFTER AN OIL PAINTING BY MICHEL JACOBS

In Root Two

IN THE WOODS AFTER AN OIL PAINTING BY MICHEL JACOBS

In overlapping Root Ones

FIG. 54. SHOWING HOW TO USE TRANSPARENT GUIDES

CHAPTER FIVE: THE WHIRLING SQUARE ROOT

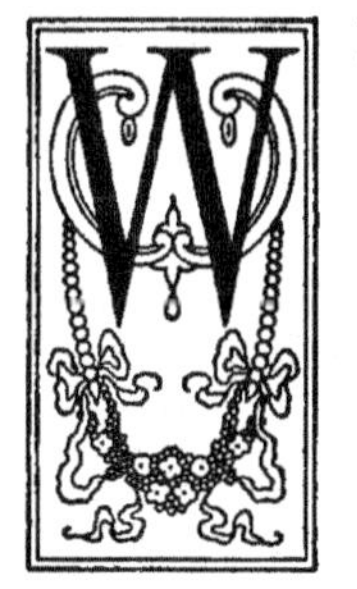

WE REMEMBER that the Whirling Square Root was made by taking the half of a square, drawing the diagonal of this half, and adding half of the square to the length of this diagonal to form one side of the oblong or rectangle. (Fig. 55.) Taking this oblong or rectangle and drawing into it a diagonal and the crossing line, and by drawing a line where the crossing line touches the outside of the oblong, we will form, as we have said before, a square on one end and a reproduction of the same form on the other end in a proportion similar to the major shape. By drawing the parallel line again, so as to form a SEQUENCE of squares, we will have produced a compositional layout in Dynamic Symmetry of the Whirling Square as is shown in Fig. 56.

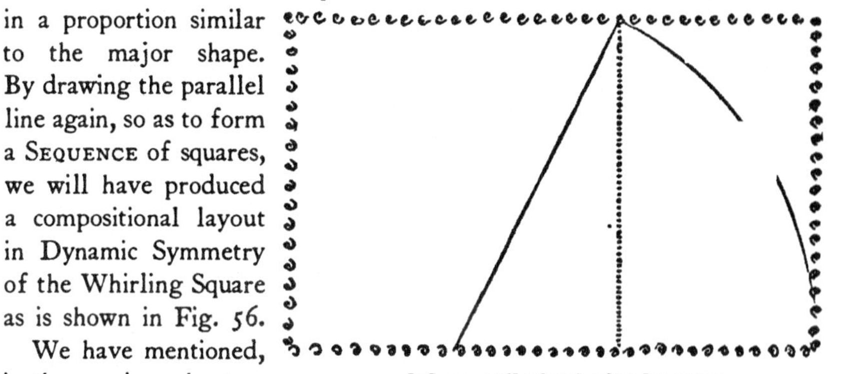

FIG. 55. WHIRLING SQUARE ROOT

We have mentioned, in the previous chapters that nature grows in an order of summation, and we have produced in Fig. 56 the Whirling Square Root in SEQUENCE. If we draw diagonal lines crossing the squares, we shall have produced a Whirling Square similar to the Greek key pattern, which is so well known. If we connect these corners with a continuous line, we shall have a spiral in the same proportion as the Greeks use on their Ionic column, as in Fig. 57.

FIG. 56. WHIRLING SQUARES IN SEQUENCE

We, of course, know the effect of centrifugal force, one of nature's phenomena. We have all seen the pinwheel throw off fire from a central point, showing us this spiral in many beautiful compositional forms.

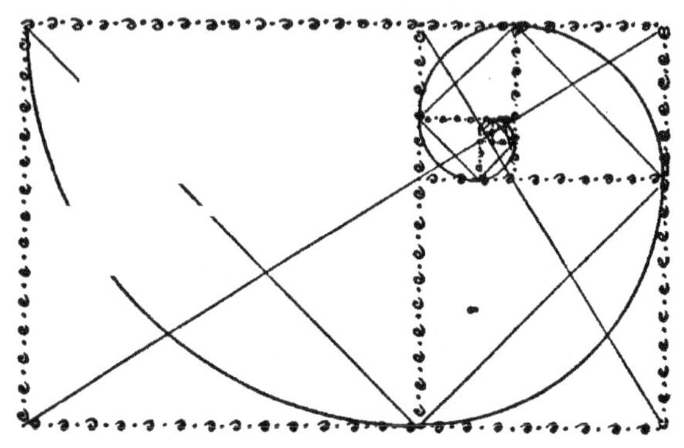

FIG. 57. WHIRLING SQUARE SHOWING GREEK KEY PATTERN

We often see this spiral reproduced in nature. Look at the illustrations in Fig. 58 for some of these.

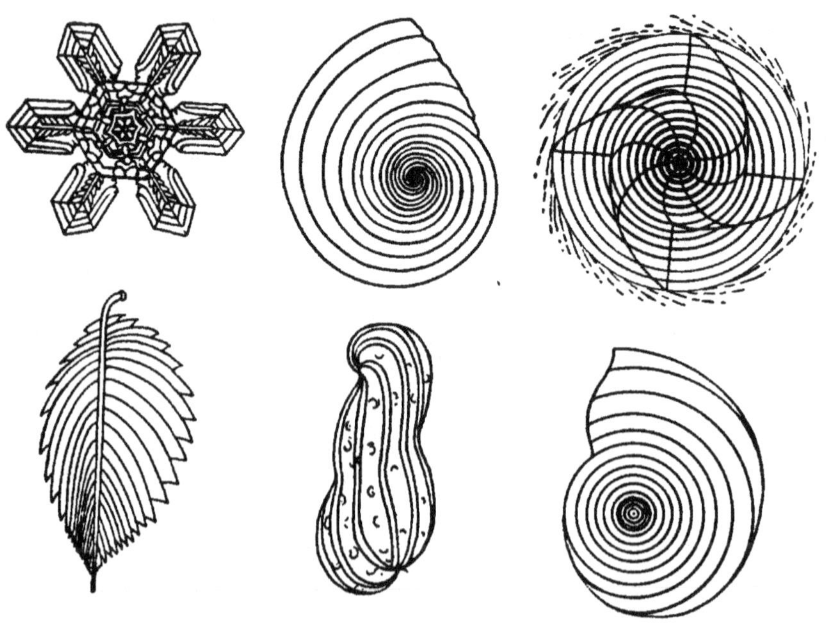

FIG. 58. NATURAL FORMS AND DESIGNS

We now wish to know how to use nature's way of making rhythmic curves which is well recognized in the scientific world and is called a logarithmic curve. Many scientists have studied this phase of nature's laws, but the artist is not interested in knowing the mathematical side of it, but simply how to reproduce it. It can be drawn *outside* of an oblong or square, but, for the present purposes, it is not necessary to do this, as at present we are only interested in producing it *inside* of a certain area.

Besides giving you these facts as to how to reproduce this Whirling Square in the Whirling Square Root, we can reproduce it in any other root. This is very important, much more important, I believe, than is generally understood by students of Dynamic Symmetry. All composition, in no matter what root, should partake somewhat of this flowing, rhythmic, compositional form: in the mind of the author.

Perhaps it is a new departure from the generally accepted ideas of Dynamic Symmetry, but to me it is a kernel of beauty, the essence of aesthetic feeling, which takes from Dynamic Symmetry the things we have seen heretofore of hard, cold, straight, angular lines. I cannot stress forcibly enough the great importance of always keeping this Whirling Square or spiral in mind.

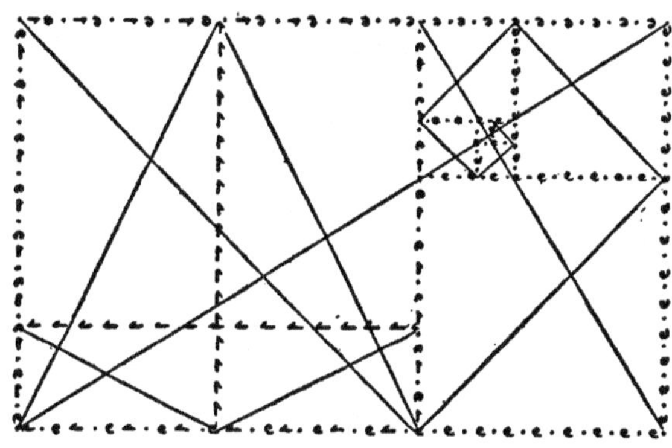

FIG. 59. WHIRLING SQUARE ROOT WITH LARGE SQUARE SUBDIVIDED

It combines and answers perfectly the well-known Hogarth lines of grace and beauty, and while in this chapter we will only take up the simpler forms, in a later chapter, it will be shown how this wonderful spiral can be used in compound forms that will help us to get rhythmic lines in beautiful proportion. One of the ways in which the Whirling Square can be used would be to take this form and draw the layout or plan, as in Fig. 57, and then to compose a picture using the eye or centre of the Whirling Square as our principal point of interest, radiating the lines more or less from this centre, and using other points as second and third points of interest, etc.

This is done by first dividing the large or first square in half, drawing the diagonal in each half, and crossing the diagonal with the crossing line and parallel line, as shown in Fig. 59. In Fig. 60 we have shown a composition

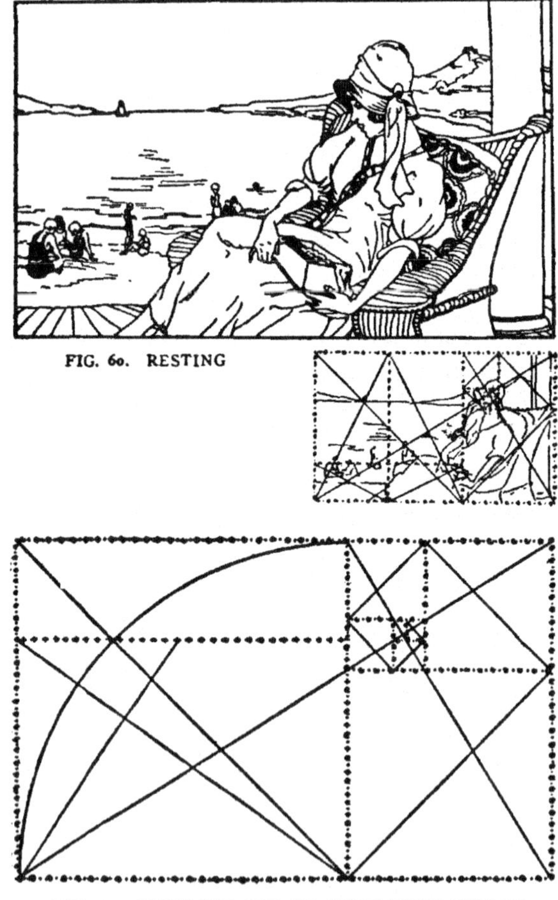

FIG. 60. RESTING

FIG. 61. WHIRLING SQUARE WITH ROOT TWO IN SEQUENCE

based on this layout. Another simple method by which the Whirling Square Root can be used would be to draw the Whirling Square, and on the large, square end, draw a diagonal making a quadrant from the two opposite corners. This will produce (as was explained in Chapter Three,

Fig. 36, and Chapter Four, Figs. 51 and 52) a Root Two inside of a square. There is shown in Fig. 61 this layout carried out.

You will notice that we have used as much of the curved lines as we thought would carry out the idea, both of the spirals and of the quarter circle or quadrant, and also the straight line.

It may now be seen that you can divide this square or the other squares in SEQUENCE into other roots to be contained in this Whirling Square Root.

In a later chapter, we will take up the proposition of making more than one Whirling Square within a Whirling Square Root.

CHAPTER SIX: SIMPLIFIED COMPOSITIONS BASED ON ROOT ONE

O PLACE a composition in a square, as we have explained in the previous chapter, you can divide the square into two Root Four's, or you can draw a diagonal and draw a quarter circle or quadrant arc finding out where the Root Two would come. Or you could use the Root Three inside of the square, or the Root Five. If you use any one of these roots in the square, you will have the balance of the square to contend with. Some of the methods of doing this are explained in the following.

Remember, it is not always necessary to follow the straight lines of the Dynamic layout, and often the curve or spiral is more pleasing, as it gives rhythm, besides placement and line. It will be understood that, of course, the spiral can be used in all roots, not alone in the Whirling Square Root, and the principals laid down in the Whirling Square Root are applicable to all roots.

Drawing a square and dividing it in half with a line drawn through the centre will give us two Root Four's. If we draw the diagonal through one of these halves and the crossing line, we shall have found out a point of interest which we will consider as the principal point of interest, and dividing the other half in the same manner to make the crossing line in the opposite corner will make the secondary point of interest. If we draw a line where the crossing line meets the boundary of one of the Root Four's and make a Root Four in SEQUENCE, we shall have produced a layout, as is shown in Fig. 62. An illustration is also shown in Fig. 63 based on this layout.

Another way of the many of using the Root One or square, would be to draw a diagonal making the quarter circle to make Root Two, as is explained in Chapter Three, Fig. 36. Taking this Root Two and drawing the diagonal, then squaring the diagonal with the short line, we shall have a point of interest which we will call the principal point. The part that is left after making Root Two we can divide in half and draw the diagonal and our crossing line to get the secondary point, as is shown in Fig. 64. In Fig. 65 is shown a composition based on this layout.

Another way that the Root One can be used would be to draw both diagonals of the square and draw two Root Two's at right angles to each other as before directed by means of two quadrant arcs or quarter circles,

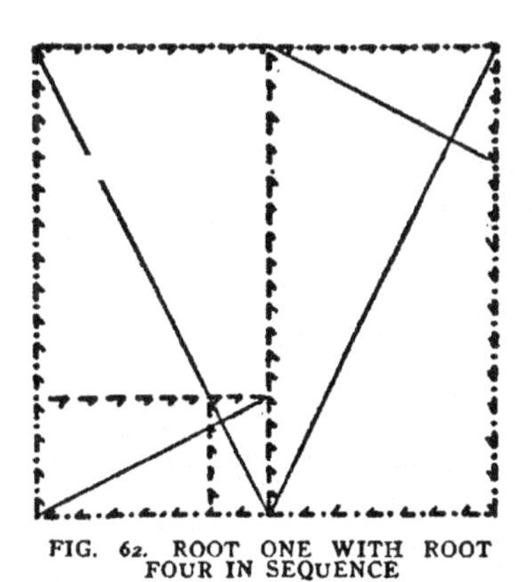

FIG. 62. ROOT ONE WITH ROOT FOUR IN SEQUENCE

FIG. 63. BARN YARD

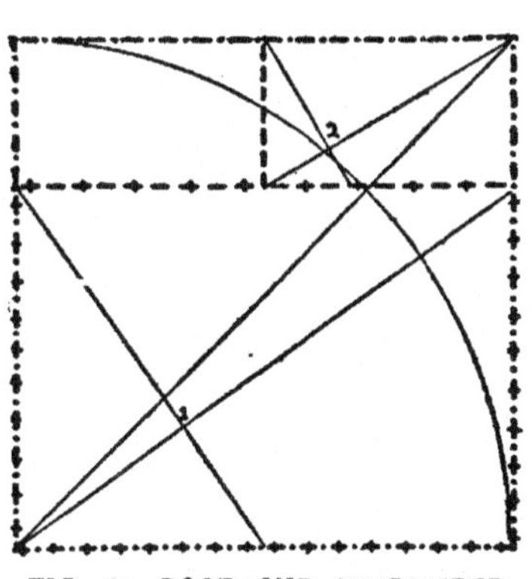

FIG. 64. ROOT ONE SUBDIVIDED INTO ROOT TWO AND TWO SMALLER ROOT THREE'S

FIG. 65. SUB-DEBS

THE GYPSY

AFTER AN OIL PAINTING BY ARTHUR SCHWIEDER

A double whirling square in a form less than Root Two

THE COURTSHIP AFTER AN OIL PAINTING BY ARTHUR SCHWIEDER

In Root Two

taking the points where the Root Two line and the quadrant arcs cross, drawing an upright line and drawing another diagonal to half of the part that was left after making the Root Two within the square, and drawing two more diagonals to half of the Root Two, as is shown in Fig. 66. In Fig. 67 is shown a composition based on this layout.

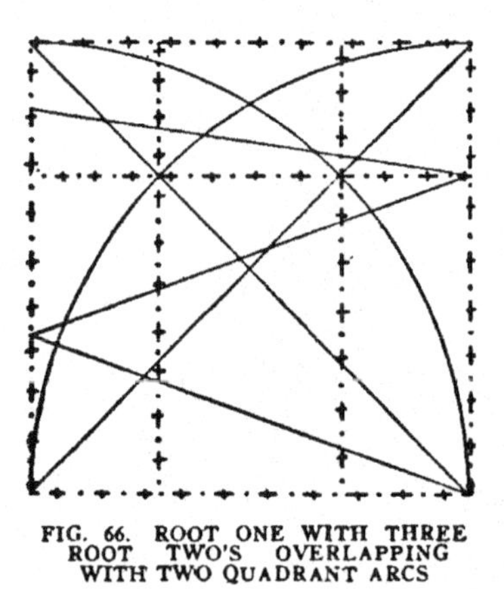

FIG. 66. ROOT ONE WITH THREE ROOT TWO'S OVERLAPPING WITH TWO QUADRANT ARCS

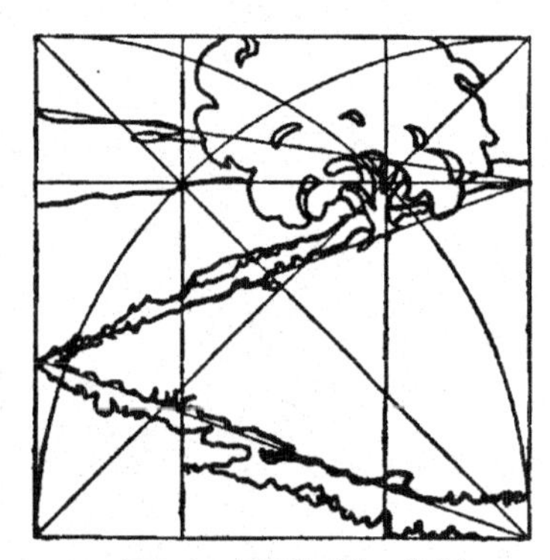

FIG. 67. TREE ON A HILL

A very beautiful conventional layout can be made in Root One by drawing four quadrant arcs and reproducing two Root Two's, as in the previous paragraph, where we make the two roots overlap, only this time we reproduce the Root Two at the side and on the top; then drawing two centre cross lines, as is shown in Fig 68, with the design carried out in Fig. 69.

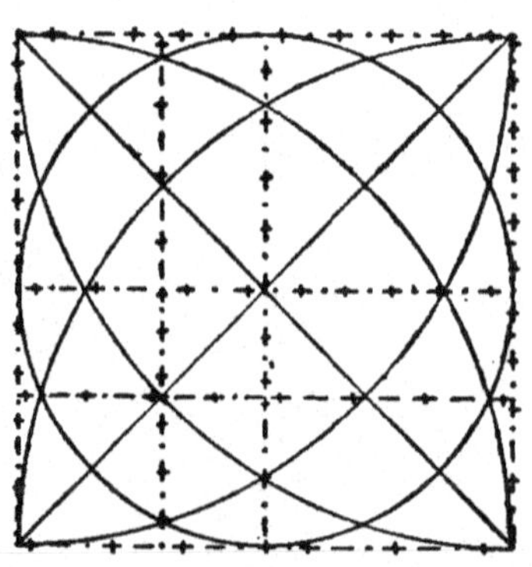

FIG. 68. ROOT ONE WITH A ROOT TWO ON TOP AND ON SIDE

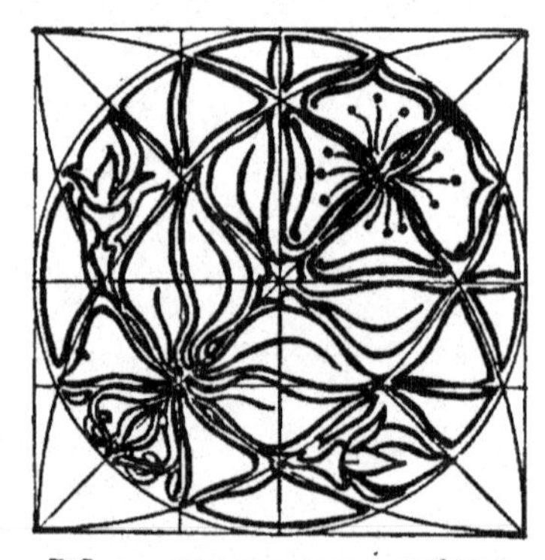

FIG. 69. CONVENTIONAL DESIGN

Many combinations using the Root One, more or less simple or complex, can be made. The more difficult ones will be explained in Chapter Thirteen.

A few divisions of a Root One will be seen on page 55.

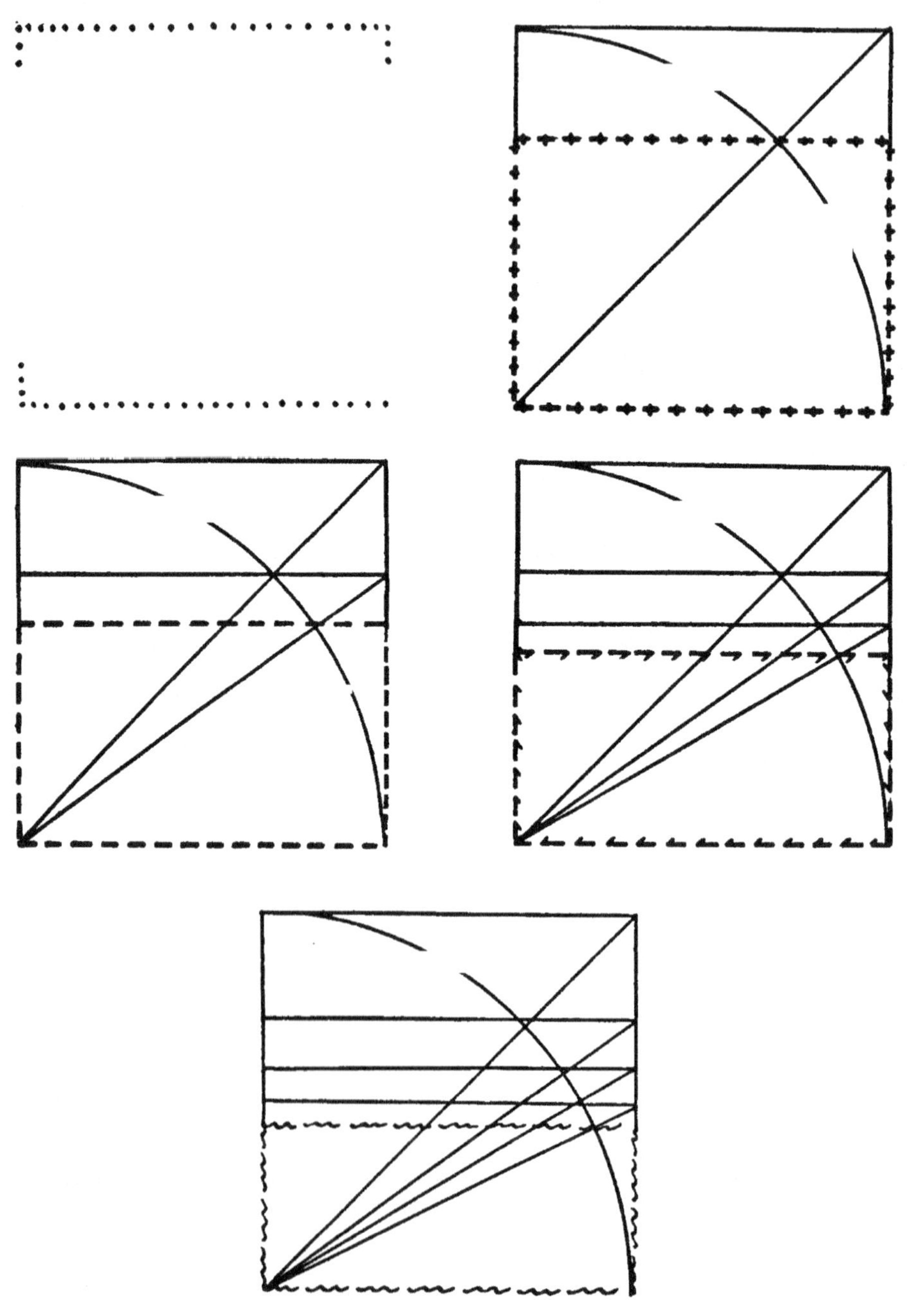

PAGE OF ROOTS TWO, THREE, FOUR, AND FIVE INSIDE ROOT ONE

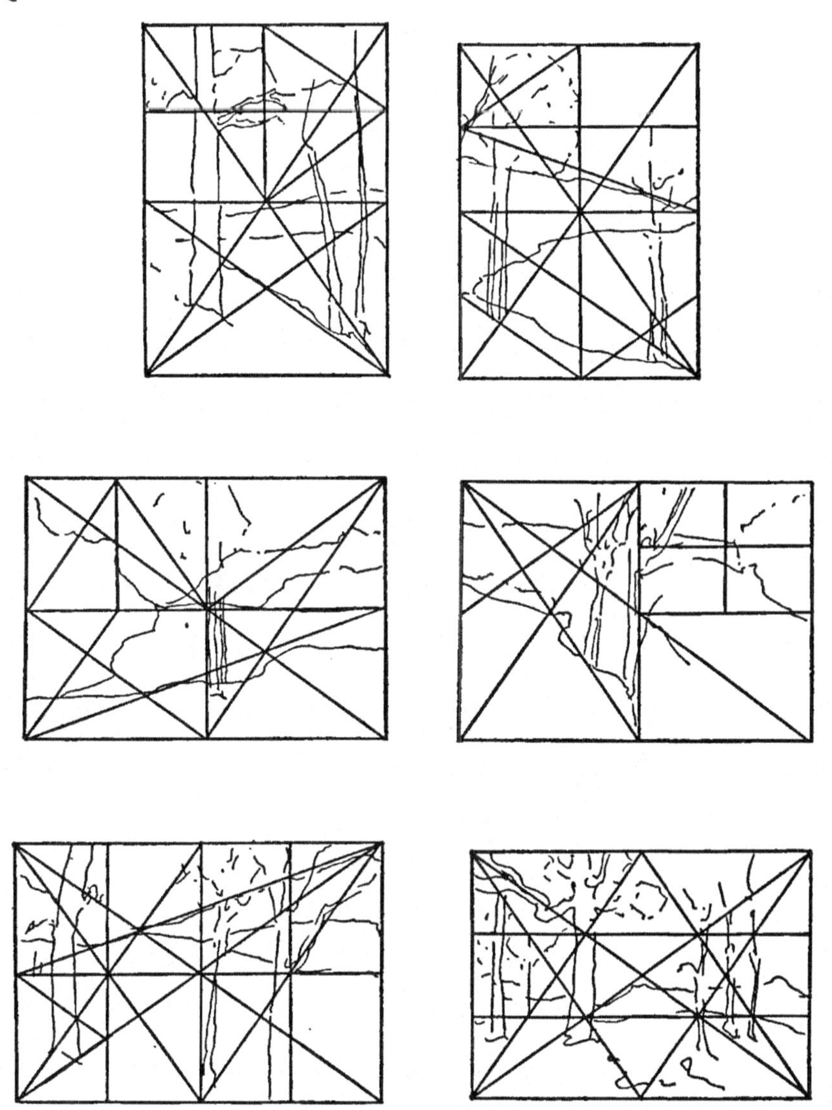

DYNAMIC LAYOUTS ALL IN ROOT TWO OF PHOTOGRAPH REPRODUCTIONS ON OPPOSITE PAGE

WINTER LANDSCAPES

By ALFRED T. FISHER

Photographed from nature in Root Two

PHOTOGRAPHS FROM NATURE By FRANK ROY FRAPRIE, S.M., F.R.P.S.

In Roots One, Two, Three, and Four

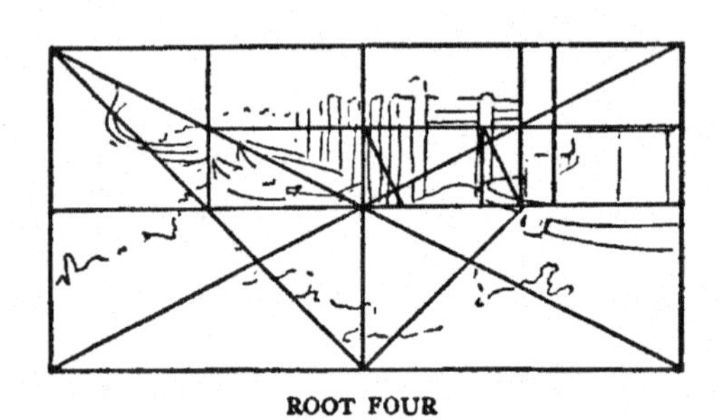

ROOT FOUR

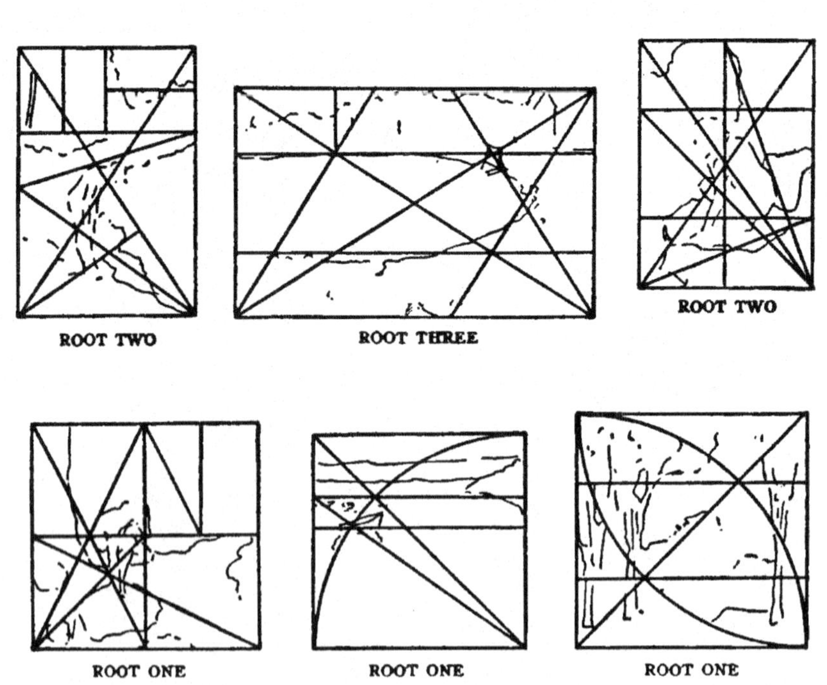

ROOT TWO ROOT THREE ROOT TWO

ROOT ONE ROOT ONE ROOT ONE

DYNAMIC LAYOUTS OF PHOTOGRAPH REPRODUCTIONS ON OPPOSITE PAGE

CHAPTER SEVEN: SIMPLIFIED COMPOSITIONS BASED ON ROOT TWO

ROOT TWO can be drawn as was explained in Chapter Two in two ways: either outside of a square or inside of a square. The best way to form Root Two or oblong outside of the square, without referring to mathematical means, would be to measure the diagonal of a square. This will give you the length of Root Two. (Illustration Fig. 29.) If you wish to know the mathematical measurement, I would refer you to Chapter Sixteen, where all the formulas are given.

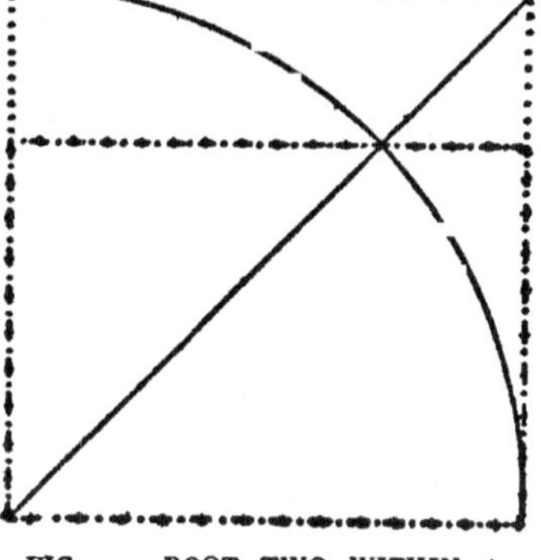

FIG. 70. ROOT TWO WITHIN A ROOT ONE

To construct Root Two inside of a square as we have explained in the previous chapters, you would draw a diagonal in the square and make the quadrant arc or quarter circle; where this line crosses the diagonal, we would draw a parallel line, as is explained in Chapter Two. This form is Root Two. The balance of the square can be thrown aside. (Fig. 70.)

For my own purposes, I have many times drawn the roots in the following way: Taking a sheet of paper of any size, and drawing the root I wished by using the square, either inside or outside method, and placing this measured root on the larger surface or canvas that I intended to use, I drew the diagonal through the opposite corners. This is one of the methods used by modern illustrators to get an enlargement. It is very simple, very practical, and does away with all mathematical means. (Fig. 71.)

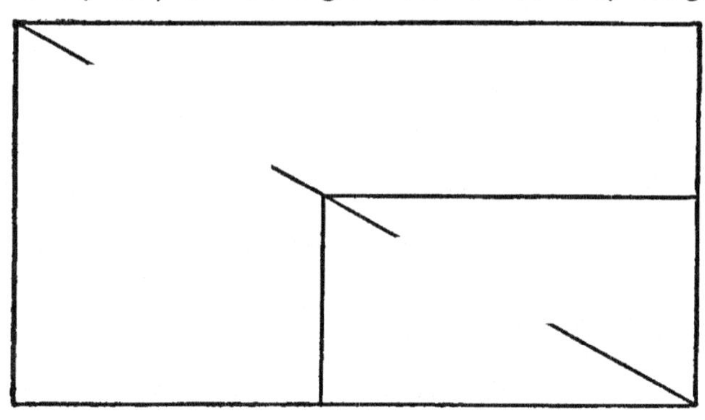

FIG. 71. ILLUSTRATIVE METHOD OF ENLARGING IN PROPORTION

After we have decided on the measurements of Root Two, we can place our composition. It is, of course, taken for granted that you have some idea to express, and that you have already conceived of the general effect that you want to get. Remember that Dynamic Symmetry is only a means of making your own conception more beautiful; it will not give you a conception, although it may suggest the arrangement. If you work for a short time with Dynamic Symmetry, your conceptions will naturally partake of compositional forms that will readily adapt themselves to some form of the innumerable Dynamic layouts.

It is for you to choose, after your first conception has been formed, which particular grid or layout you wish to use; which particular thing you wish to stress and make the principal point of interest, which the second, third, etc. For example, you may be drawing or photographing a landscape, you may have houses and trees and lakes, any one of these three objects could be placed in the principal point of interest, depending on what your original conception of the subject was or how the ladscape impressed you. This is the artists' feeling which is individual, and is, of course, real art.

Many students of Dynamic Symmetry have been led to believe that by drawing the lines of Dynamic Symmetry they can get an idea or conception. This, in my mind, is a mistake. The conception must come first; then use Dynamic Symmetry to perfect the arrangement.

Taking the Root Two, we can use many simple forms or layouts, one of which would be to draw the diagonal, drawing the squaring line through the diagonal, and then dividing the remaining part in the same manner, as is illustrated in Fig. 72. After we have made the two Root Two's, one on each side of the large Root Two (this root divides exactly into two Root Two's) we can again draw the diagonal in the smaller ones, finding out the points of interests in each. One of these may be selected as the principal point of interest and the other as the secondary. The remaining parts of the smaller Root Two's we can divide into two equal parts (again two Root Two's) as is shown in Fig. 73. In Fig. 74 is shown an illustration based on this compositional layout.

Root Two divides itself exactly into two equal Root Two's by means of drawing the diagonal and the crossing lines, as is shown in Fig. 75.

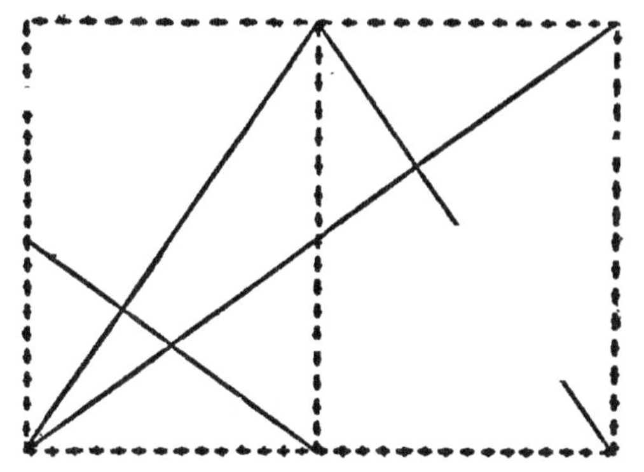

FIG. 72. THE DIAGONAL AND CROSSING LINE IN ROOT TWO AND THE ROOT TWO IN SEQUENCE

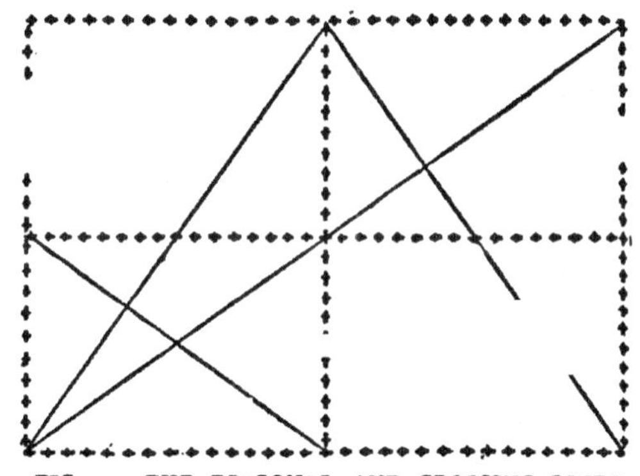

FIG. 73. THE DIAGONAL AND CROSSING LINES IN ROOT TWO AND THE TWO ROOT TWO'S IN SEQUENCE

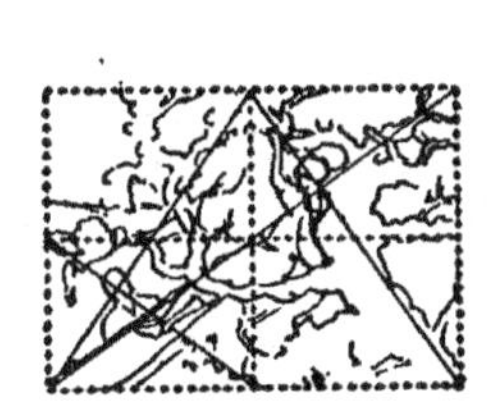

FIG. 74. THE GARDEN WALL

Drawing two diagonals and making the star layout by crossing the diagonals in both places, and then by drawing a line through the intersection and a parallel line through the length of the oblong, you will have made eight divisions which will all be Root Two's, and the two halves will also be two Root Two's in greater SEQUENCE; and again, you will find four Root Two's, so that this layout really gives you one large Root Two which contains two Root Two's. These smaller Root Two's also contain four Root Two's. The major form contains also eight Root Two's, as is shown in Fig. 76.

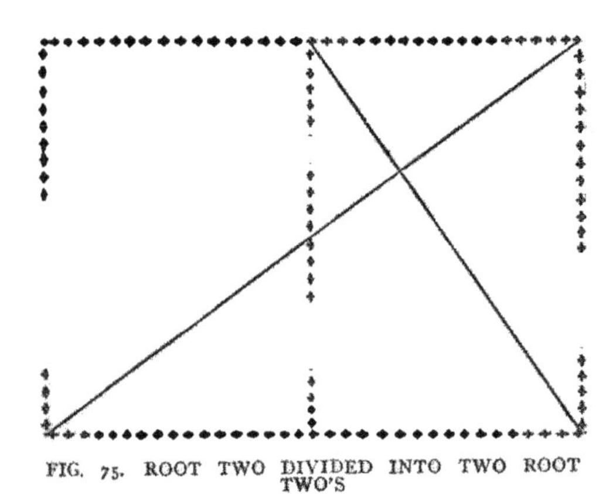

FIG. 75. ROOT TWO DIVIDED INTO TWO ROOT TWO'S

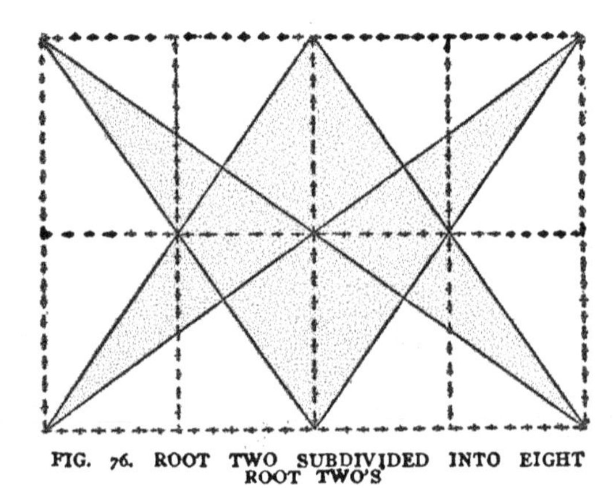

FIG. 76. ROOT TWO SUBDIVIDED INTO EIGHT ROOT TWO'S

Another method of using a simple composition inside of a Root Two would be to form a Root One or square at the side, and in the remaining part to draw again a square in the smaller proportion, or SEQUENCE. This would make a small square and a Root Two in smaller proportion of SEQUENCE. If we draw a quadrant arc or a quarter circle from the far corners of the large square and cross this with the diagonal line, we shall have found the place where again the Root Two will be inside the square. If we draw the diagonal and crossing lines in the small Root Two, we shall have the point which we can make one of our points of interests. This layout would then look like Fig. 77. In Fig. 78 is shown an illustration based on this layout.

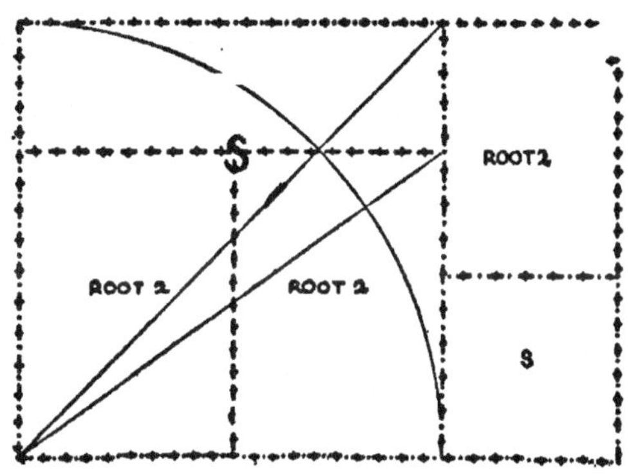

FIG. 77. ROOT TWO WITH ROOTS ONE AND TWO IN SEQUENCE

FIG. 78. LA TOILETTE

Another very simple composition inside of a Root Two would be to make a square, as we did in the previous layout, but this time make Root One's on both sides, one overlapping the other. This would then look like Fig. 79. If we take three of the Root Two's thus formed and draw the diagonal with a crossing line and draw the diagonals in three of the squares that remain, we will have a layout which will look like Fig. 80. In Fig. 81 is shown a composition based on this layout.

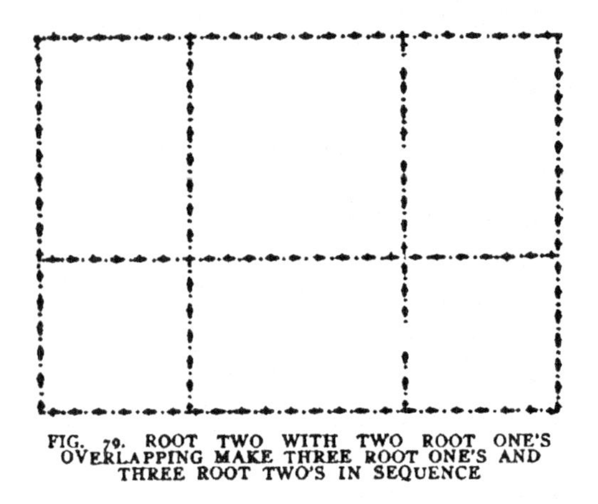

FIG. 79. ROOT TWO WITH TWO ROOT ONE'S OVERLAPPING MAKE THREE ROOT ONE'S AND THREE ROOT TWO'S IN SEQUENCE

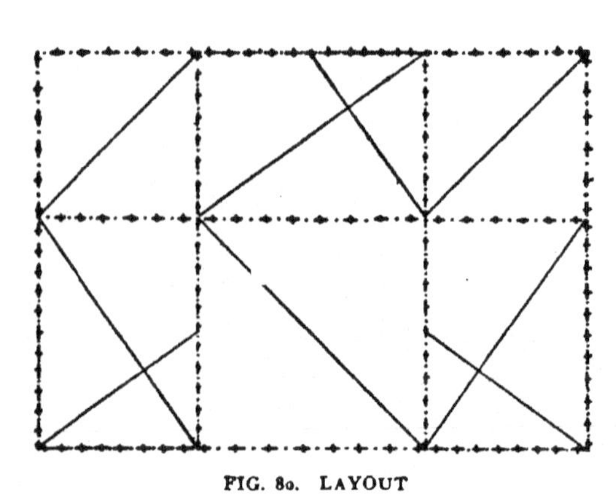

FIG. 80. LAYOUT

FIG. 81. THE SHOP WINDOW

Another simple method of using a Root Two would be to draw a square within the Root Two and in this square draw two Root Two's overlapping each other. This would give you the layout shown in Fig. 82. In Fig. 83 is shown a composition based on this layout.

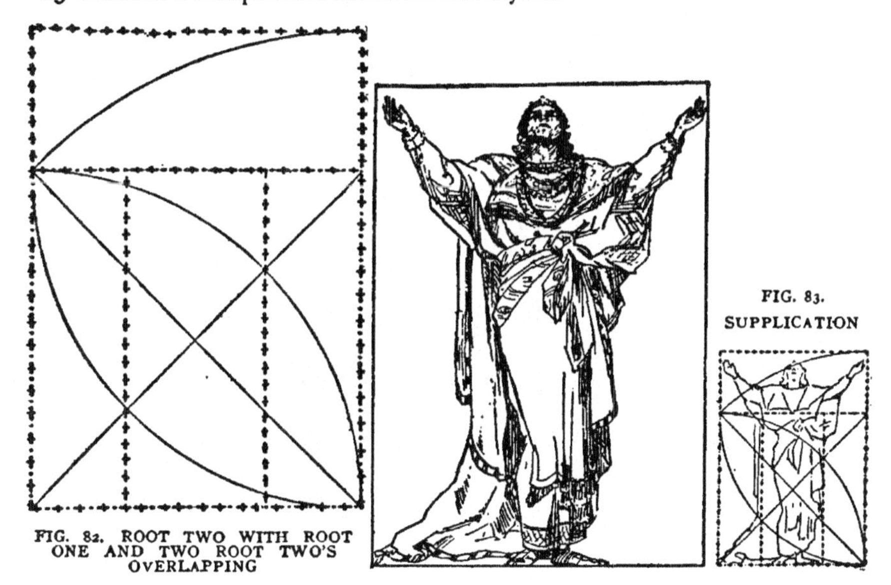

FIG. 82. ROOT TWO WITH ROOT ONE AND TWO ROOT TWO'S OVERLAPPING

FIG. 83. SUPPLICATION

Another compositional layout can be made by drawing the square on one end of the Root Two, and dividing the remaining part into a square and a Root Two, and making the principal point of interest on the centre of the diagonals of the square and the secondary point of interest in the smaller square, as is shown in Fig. 84. There is shown in Fig. 85 a composition based on this layout. On page 70 are shown a few layouts in Root Two.

You can readily understand by these that the combinations are inexhaustible. In Chapters Thirteen and Fourteen, there is shown the combination of putting other roots and more complex compositions within the Root Two.

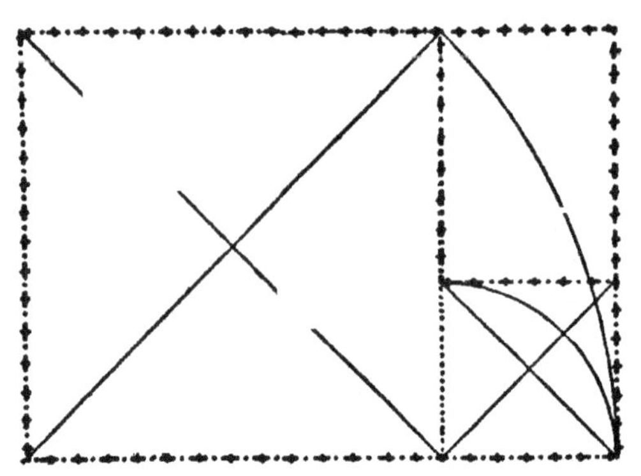

FIG. 84. ROOT TWO WITH ROOT ONE ON SIDE FORMING ROOT TWO AND ROOT ONE ON END

FIG. 85. AFTER THE SNOWSTORM

COMMERCIAL COMPOSITIONS IN ROOT TWO

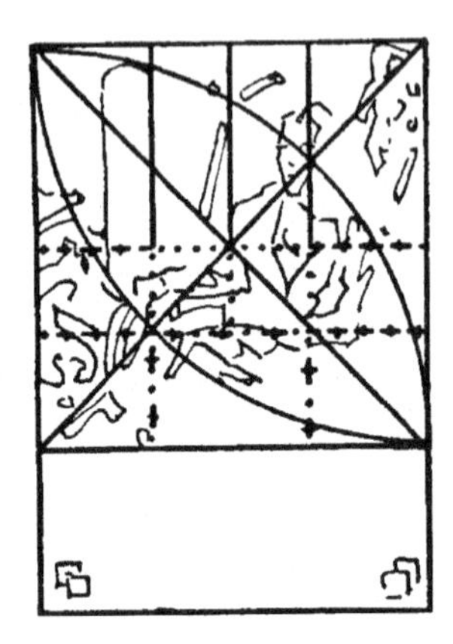

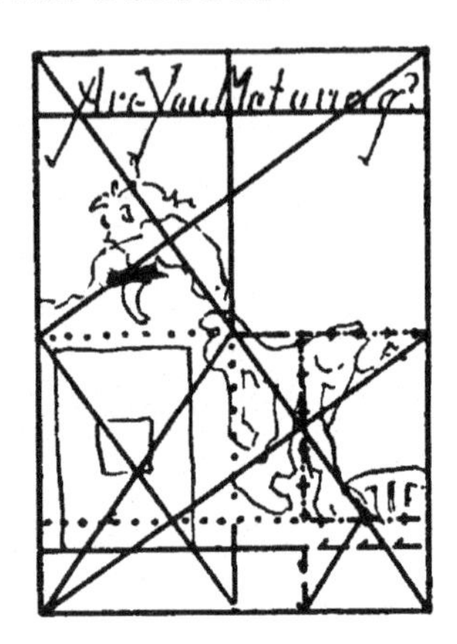

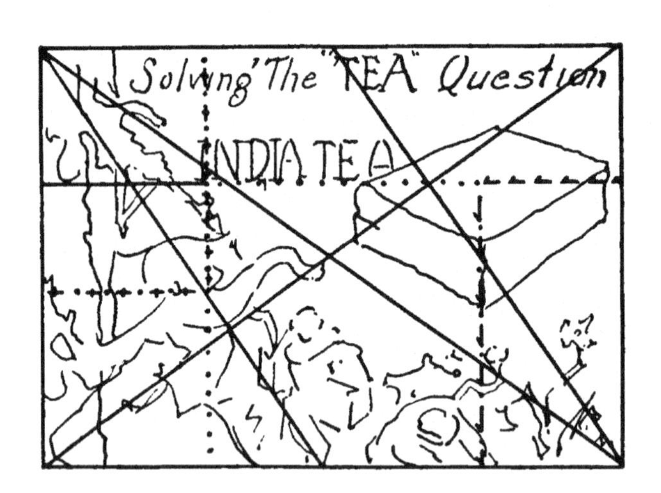
Solving The "TEA" Question
INDIA TEA

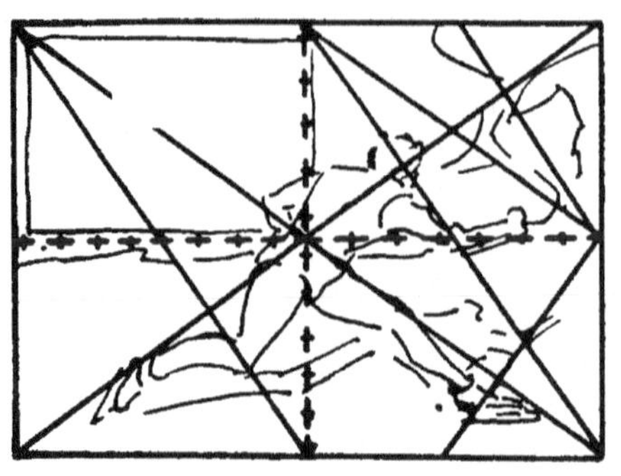

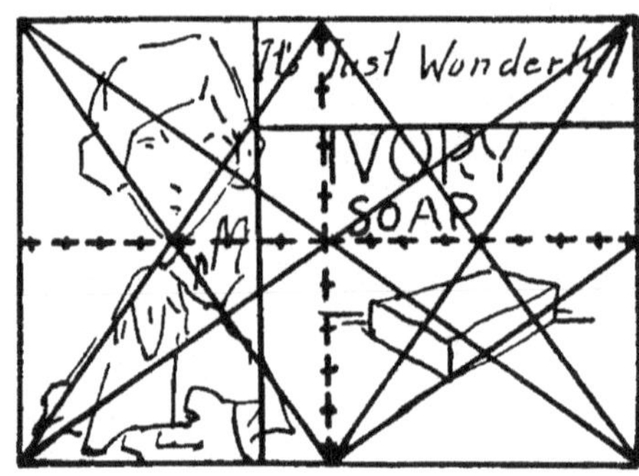
IVORY
SOAP

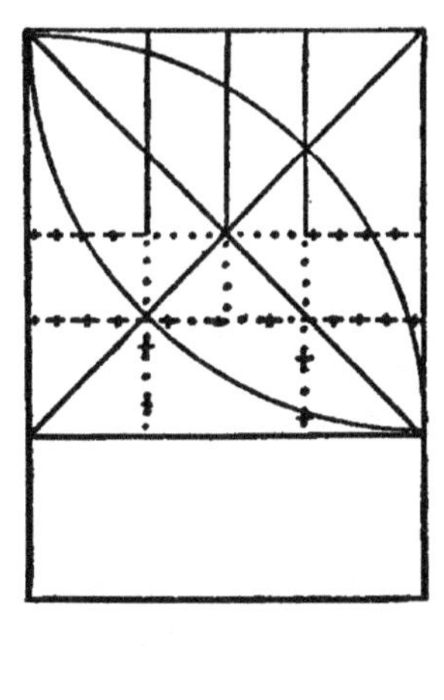

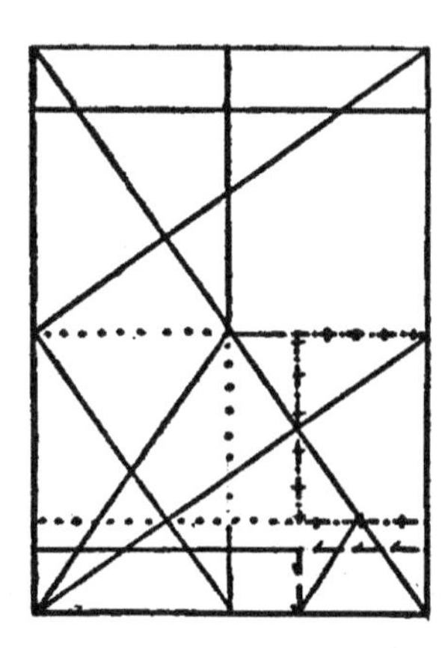

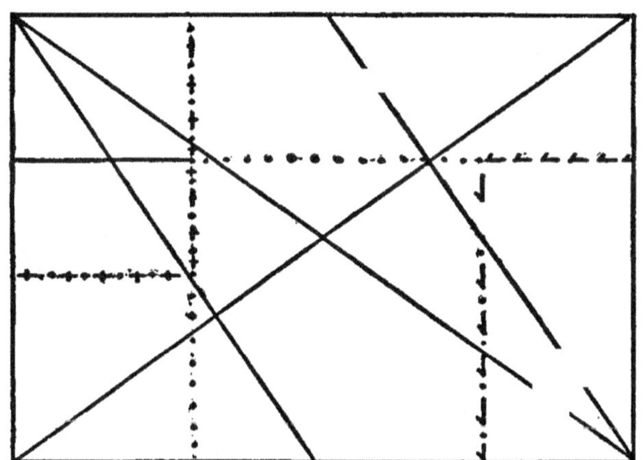

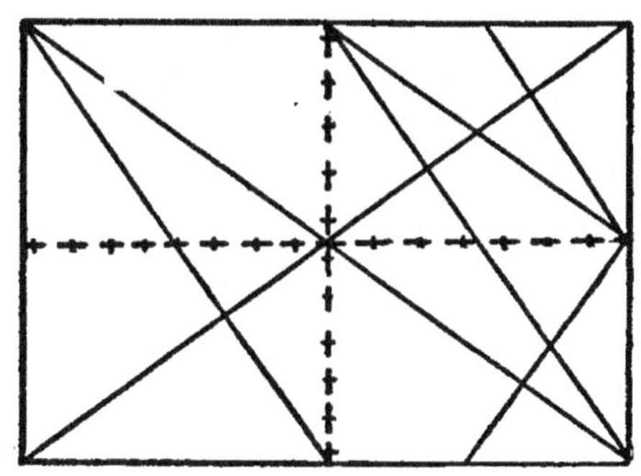

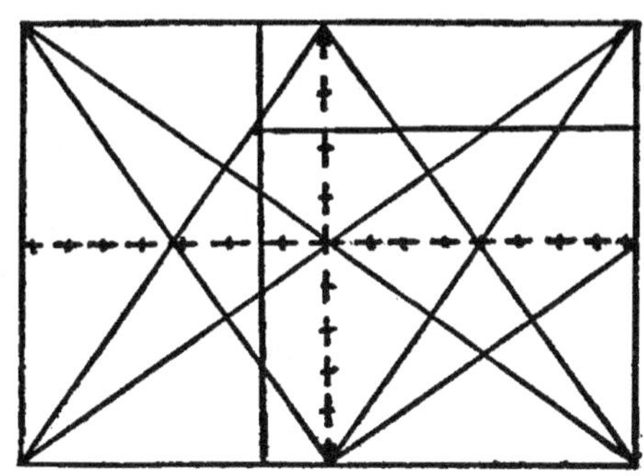

LAYOUTS BASED ON PAGE 69

CHAPTER EIGHT: ROOT THREE

THIS root is found by the different methods shown in Chapter Two, either outside of the square or inside of the square. If it is necessary to find out the size of a canvas desired, it can be done by means of the small size root as is explained in Chapter Seven and drawing the diagonal through the corners to get a larger proportion of the same root—the same method that an illustrator uses to enlarge his sketch, as is illustrated in Fig. 71. One can also find the root desired by using the transparent layouts, as before described.

Taking a Root Three, draw the diagonals and the squaring lines on both ends, crossing both diagonals. This will give you three Root Three's. On the lower Root Three draw two uprights where the diagonal crosses the parallel line. On the upper Root Three, divide in half and draw two diagonals to meet at the centre, as is shown in Fig. 86. In Fig. 87 there is shown a composition based on this layout.

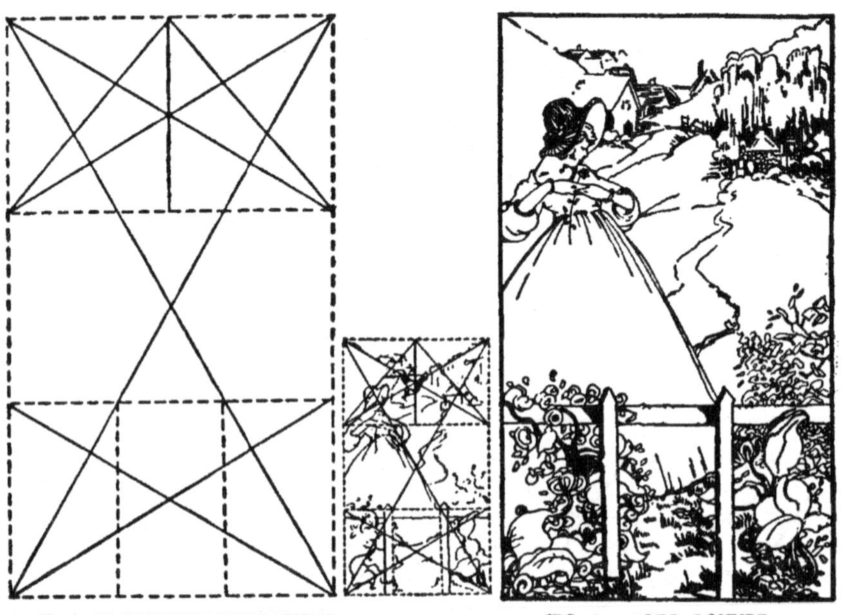

FIG. 86. ROOT THREE WITH THREE ROOT THREE'S IN SEQUENCE

FIG. 87. POKE BONNET

Another compositional layout in Root Three would be to draw the two diagonals and the two crossing lines (as in the previous example), making three Root Three's. In the lower Root Three, draw a Whirling Square, as is shown in Fig. 88. In Fig. 89 there is shown a composition based on this plan.

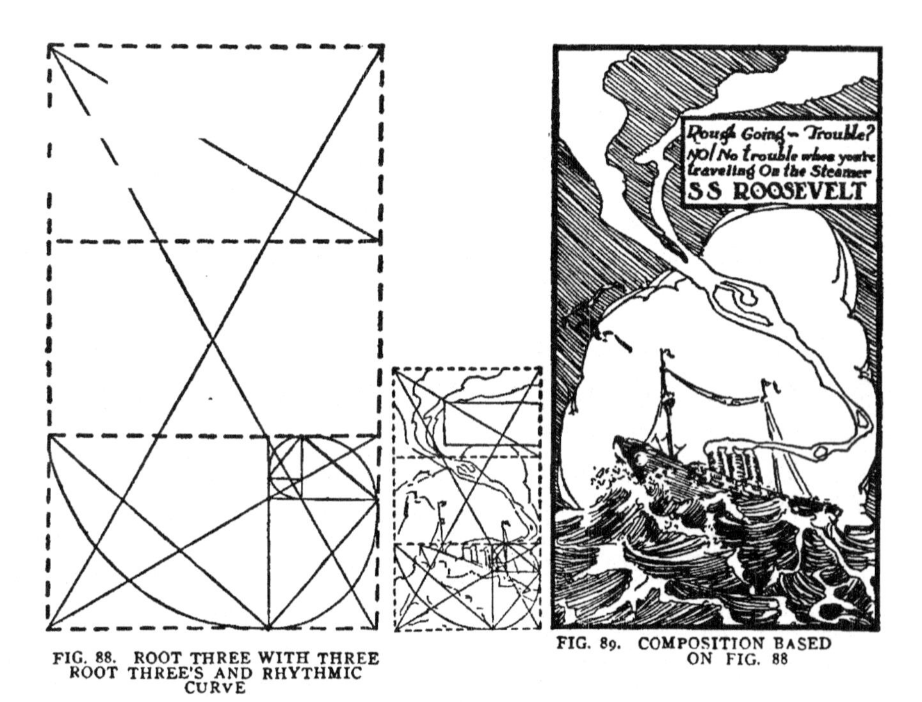

FIG. 88. ROOT THREE WITH THREE ROOT THREE'S AND RHYTHMIC CURVE

FIG. 89. COMPOSITION BASED ON FIG. 88

Still another method of laying out Root Three would be to draw a square on one side and a square on the opposite side. These will overlap each other at the centre. In other words, you would have two Root One's overlapping in a Root Three. Draw both diagonals in both squares, and in one square reproduce Root Three on the right-hand side, as is shown in Fig. 90. In Fig. 91 is shown a composition based on this layout.

It can be readily understood that the number of plans or layouts are too numerous to mention. The foregoing is only to give you some idea of some of the layouts.

On page 74 there is shown a number of layouts in Root Three.

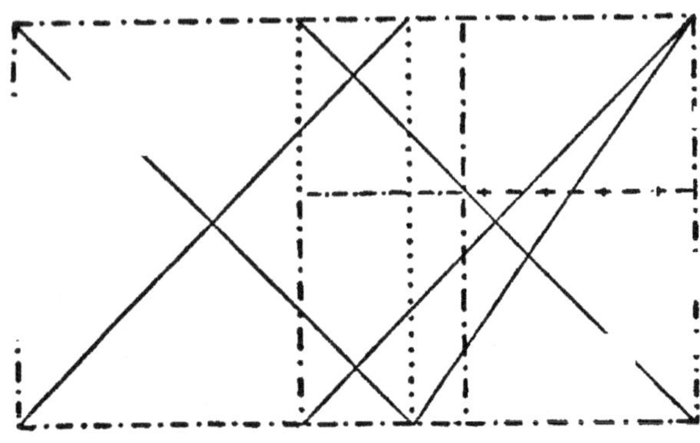

FIG. 90. ROOT THREE WITH OVERLAPPING ROOT ONE'S AND ROOT THREE IN SEQUENCE

THE ART OF COLOR

In these days of technical knowledge and scientific accuracy, It is a great wonder that the artist still follows the old law of colors and their complementaries as demonstrated by Newton and Brewster, based on the theory that red, blue, and yellow are primary colors, and green, purple, and orange are secondary. The theory has long since been discarded by scientists and the new theory adopted as laid down by Young-Helmholtz-Tyndall on the spectrum

When we see an object that is a certain color in a white light the shadows of that object assume a color that is toward the complementary to the color of the lighted side.

CONTENTS

LIST OF ILLUSTRATIONS

FIG. 91. COMPOSITION BASED ON FIG. 90

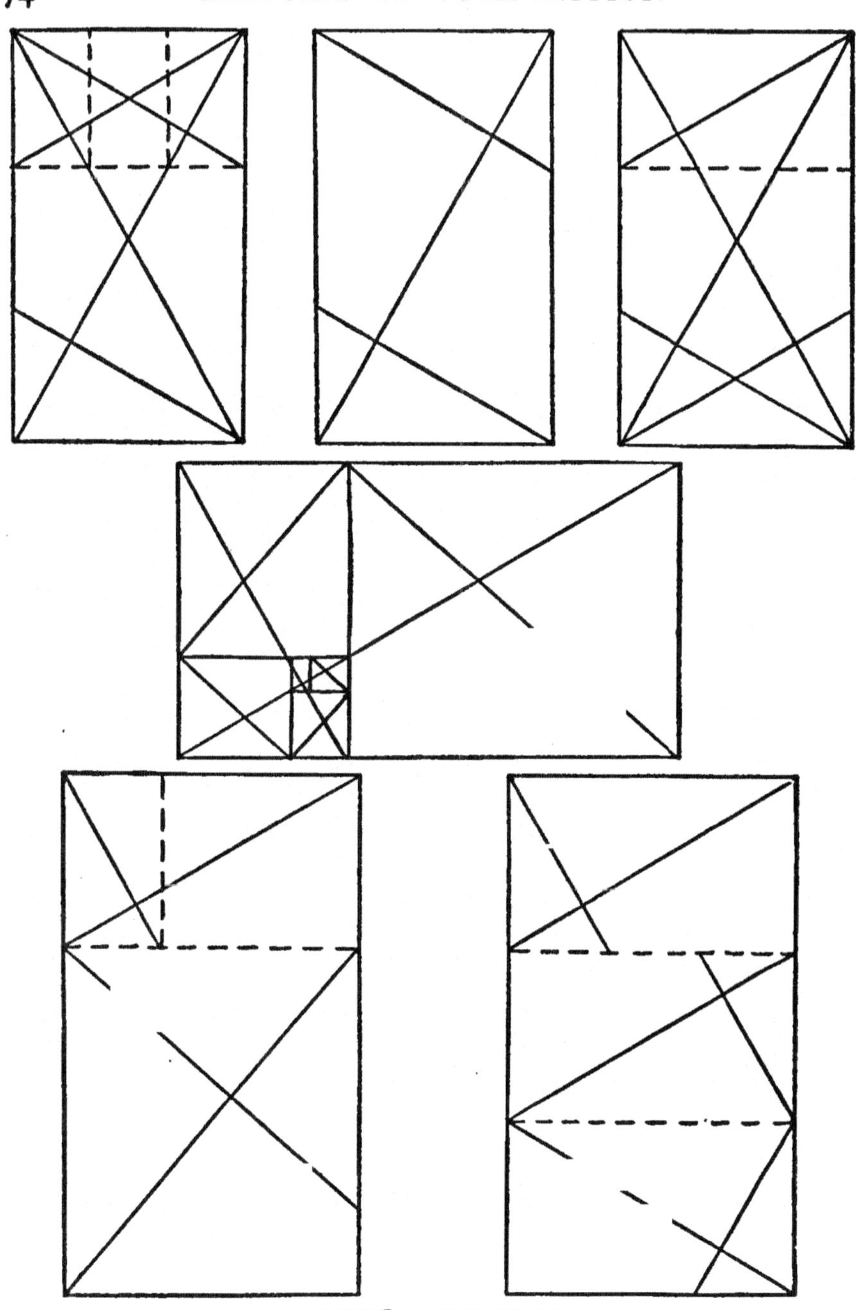

LAYOUTS IN ROOT THREE

PHOTOGRAPHED FROM NATURE

By FRANK ROY FRAPRIE, S. M., F R P. S.

In Root Two

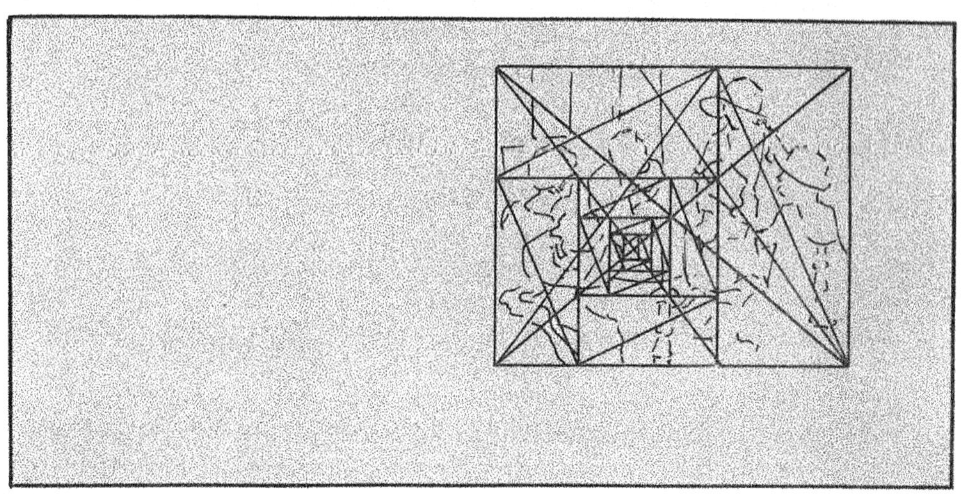

THE WILL — AFTER AN OIL PAINTING BY ARTHUR SCHWIEDER

A form less than Root Two using whirling square motif and star layout

CHAPTER NINE: ROOT FOUR

OOT FOUR is two squares laid alongside of each other, as is shown in Fig. 92. If we take this Root Four and draw the diagonals and the squaring lines and the parallel lines to the base, we shall have made two squares and four Root Four's inside of this large oblong, as is shown in Fig. 93. There is shown an illustration in Fig. 94 based on this plan.

Another interesting compositional layout can be made by taking the two Root One's which Root Four contains, and dividing them with each of the diagonals and squaring the diagonals, and on both of these diagonals drawing the quadrant arc or quarter circle, which will produce two Root Two's in each square, one overlapping the other, as is shown in Fig. 94. There is shown a composition in Fig. 96 based on this layout.

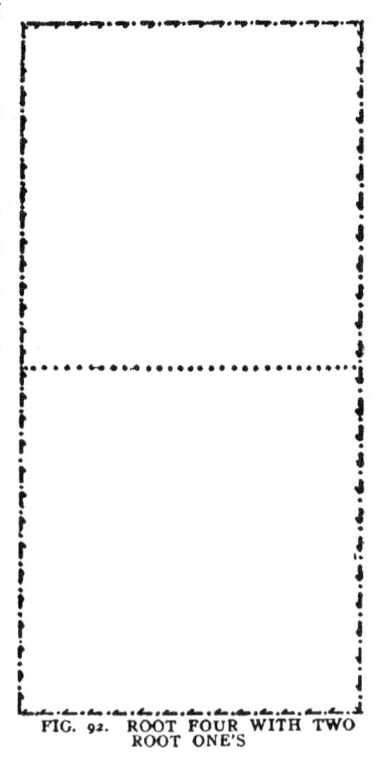

FIG. 92. ROOT FOUR WITH TWO ROOT ONE'S

Another compositional form would be to take the Root Four and lay off the Whirling Square Root by taking the compositional layout in Fig. 93, which will give half of the square. The crossing line, in that instance, would give the length of the Whirling Square if added to half of the square, as was explained in Chapter Five on the Whirling Square Root. This will give, inside of the Root Four, a large and a smaller Whirling Square Root and two Root One's. If we divide this Whirling Square Root by a diagonal and the crossing line completing the smaller form in SEQUENCE, we will have a compositional layout, as is shown in Fig. 97. In Fig. 98 there is shown a composition based on this layout.

Numbers of compositional layouts of Root Four are shown on page 81, so that it will not be necessary to give a separate explanation of each.

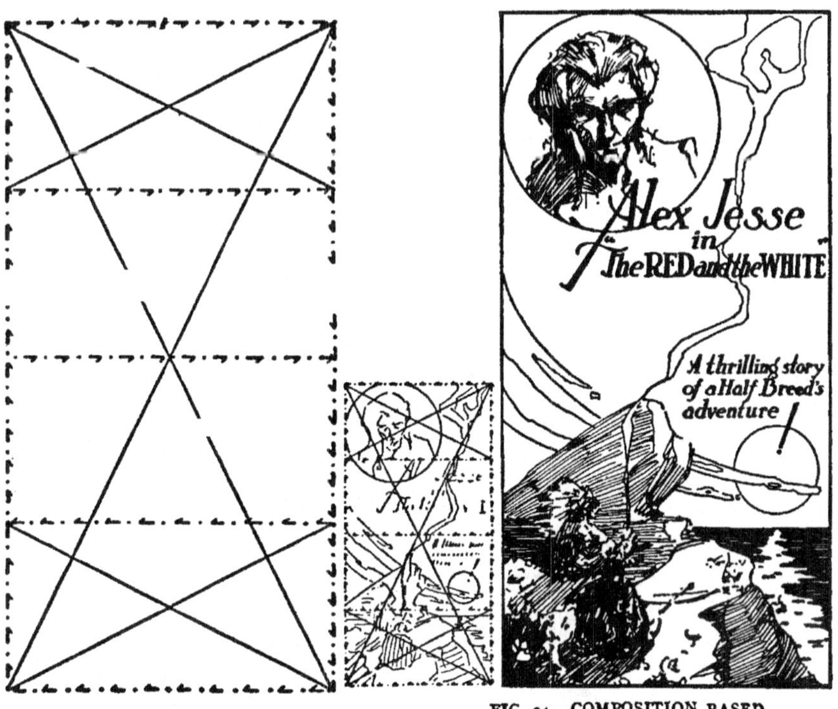

FIG. 93. ROOT FOUR WITH TWO ROOT ONE'S AND FOUR ROOT FOUR'S

FIG. 94. COMPOSITION BASED ON FIG. 93

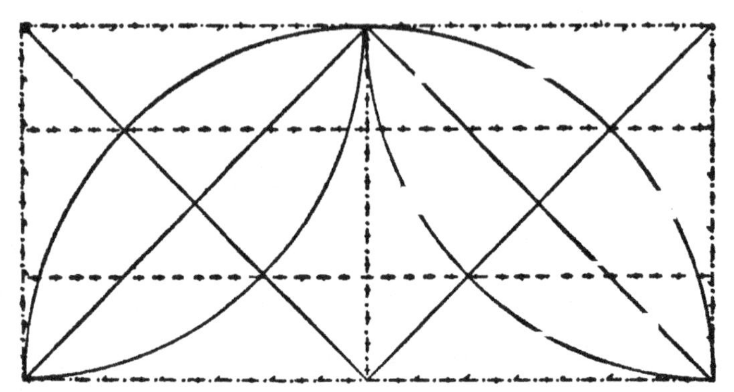

FIG. 95. ROOT FOUR WITH TWO ROOT ONE'S, EACH SQUARE CONTAINING TWO ROOT TWO'S OVERLAPPING

FIG. 96. COMPOSITION BASED ON FIG. 95

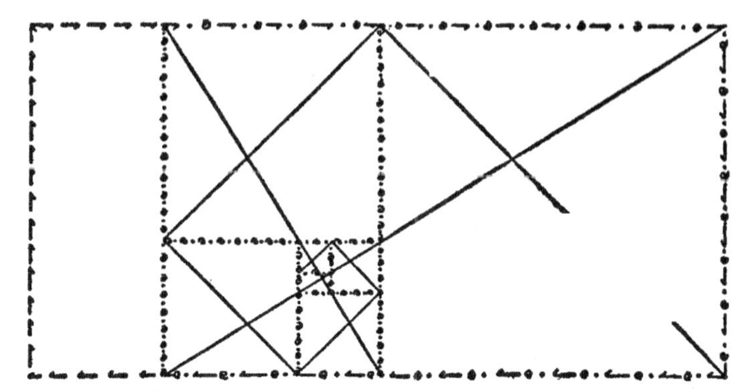

FIG. 97. ROOT FOUR WITH WHIRLING SQUARES IN SEQUENCE AND TWO ROOT ONE'S

FIG. 98. COMPOSITION BASED ON FIG. 97

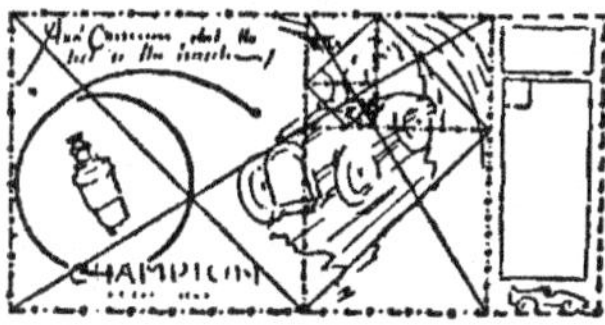

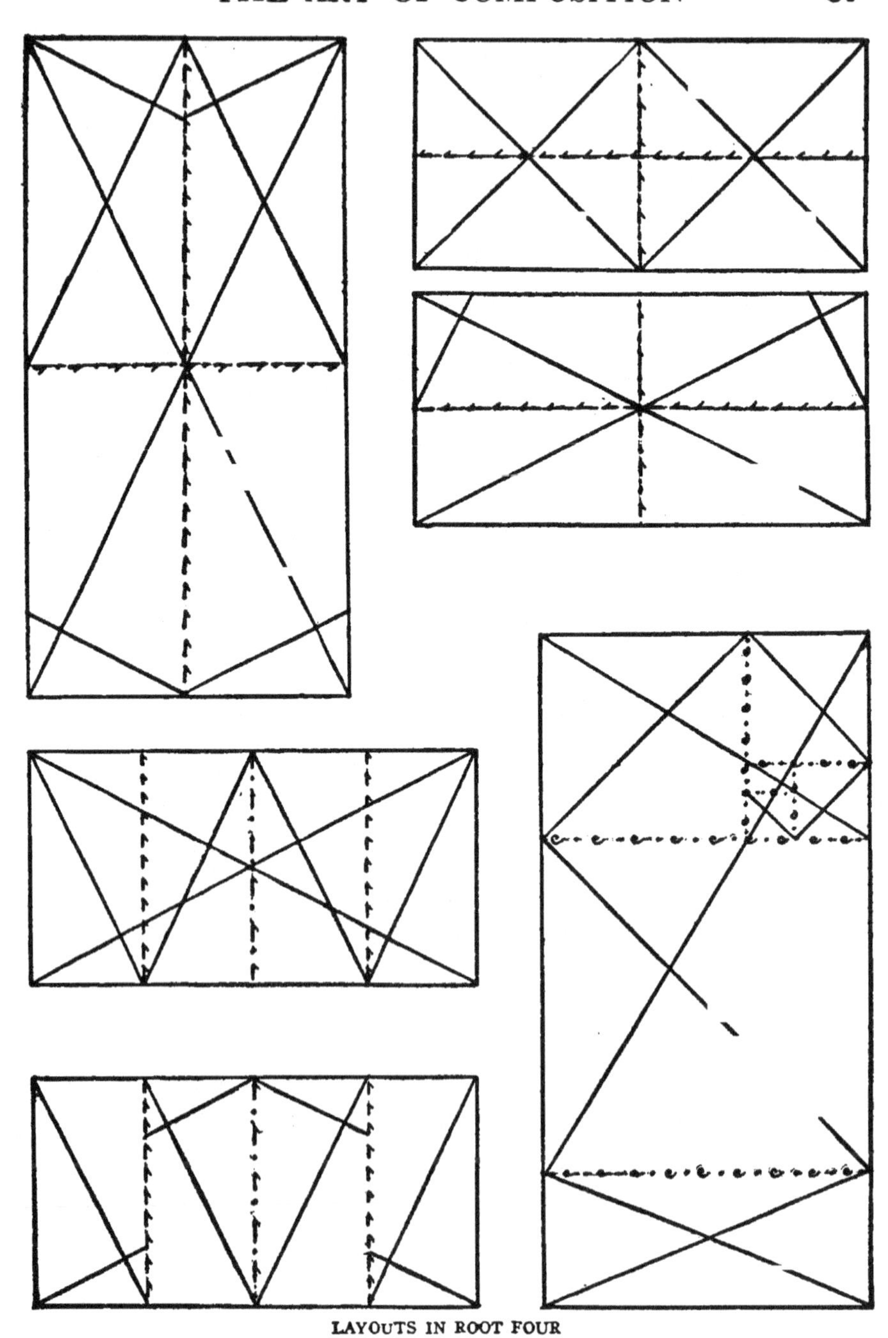

LAYOUTS IN ROOT FOUR

CHAPTER TEN: ROOT FIVE

THE Root Five oblong or rectangle is, perhaps, one of the most interesting of all of the roots. It is the one, I believe, on which the whole of Dynamic Symmetry is based. Some of the forms in SEQUENCE into which it can be divided are as follows:

A Root Five can contain two Whirling Squares and two Root Fours. (Fig. 99.)

It can be divided into one Whirling Square horizontal and one Whirling Square perpendicular. (Fig. 100.)

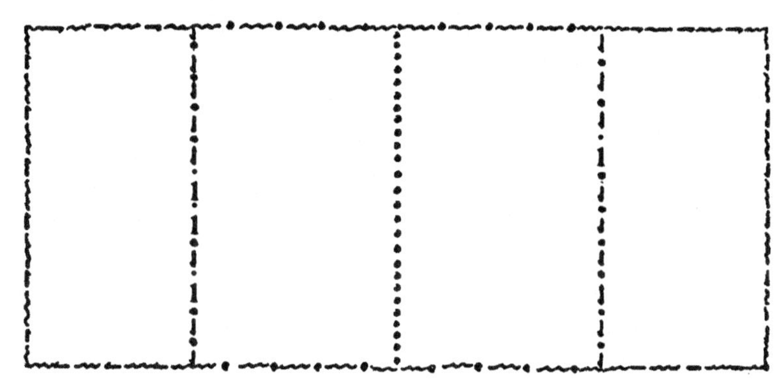

FIG. 99. ROOT FIVE CONTAINING TWO ROOT FOUR'S AND TWO WHIRLING SQUARES

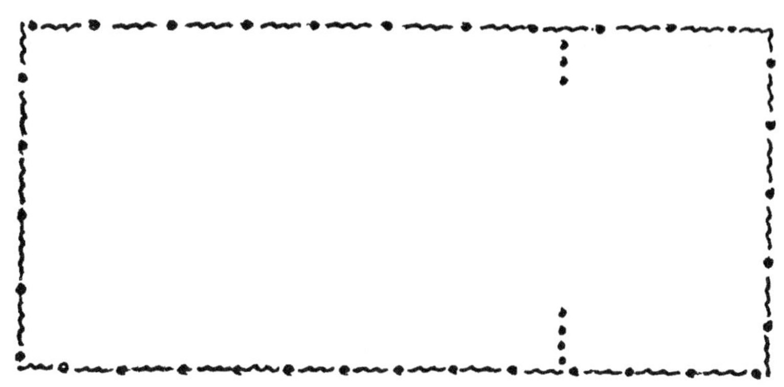

FIG. 100. ROOT FIVE CONTAINING A HORIZONTAL AND PERPENDICULAR WHIRLING SQUARE

A Root Five rectangle can be divided into a square and two Whirling Square Roots. (Fig. 34.)

Many more of these combinations with other roots can be made with a Root Five. This root is also peculiar, as it will give us the dimensions of the human figure based on the idea that the human figure can be enclosed in two Root Five's laid alongside each other.

The proportions of the human figure have been explained by Mr. Jay Hambidge in the *Diagonal* so thoroughly that I must refer the reader again to "The Elements of Dynamic Symmetry" for this information, as this book is intended only as a book of composition.

The student, for years, has been taught at first to draw from the antique, and until the advent of Dynamic Symmetry, we considered the figures of the Greeks to be taken from life, but we now know that they were conventionalizations based on the ideal, using Dynamic Symmetry to get the placement of the different members of the human body. From my experience, I find that it is much better to start the student drawing from life, and later on, after he has learned Dynamic Symmetry, to apply his life drawing and anatomy, together with Dynamic Symmetry measurements, to the perfection of his design.

On page 84 you will see a number of layouts using some of the different divisions of Root Five.

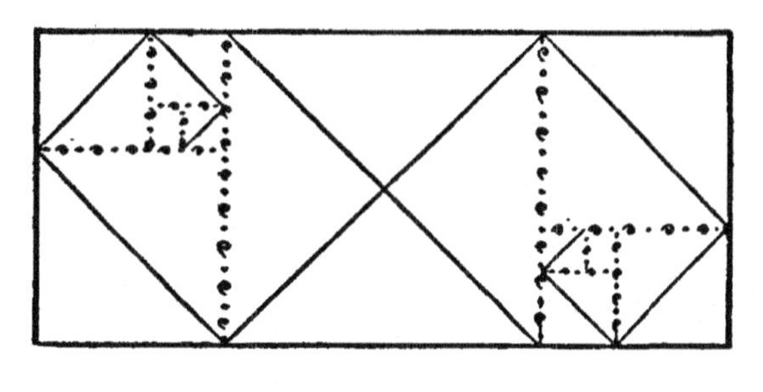

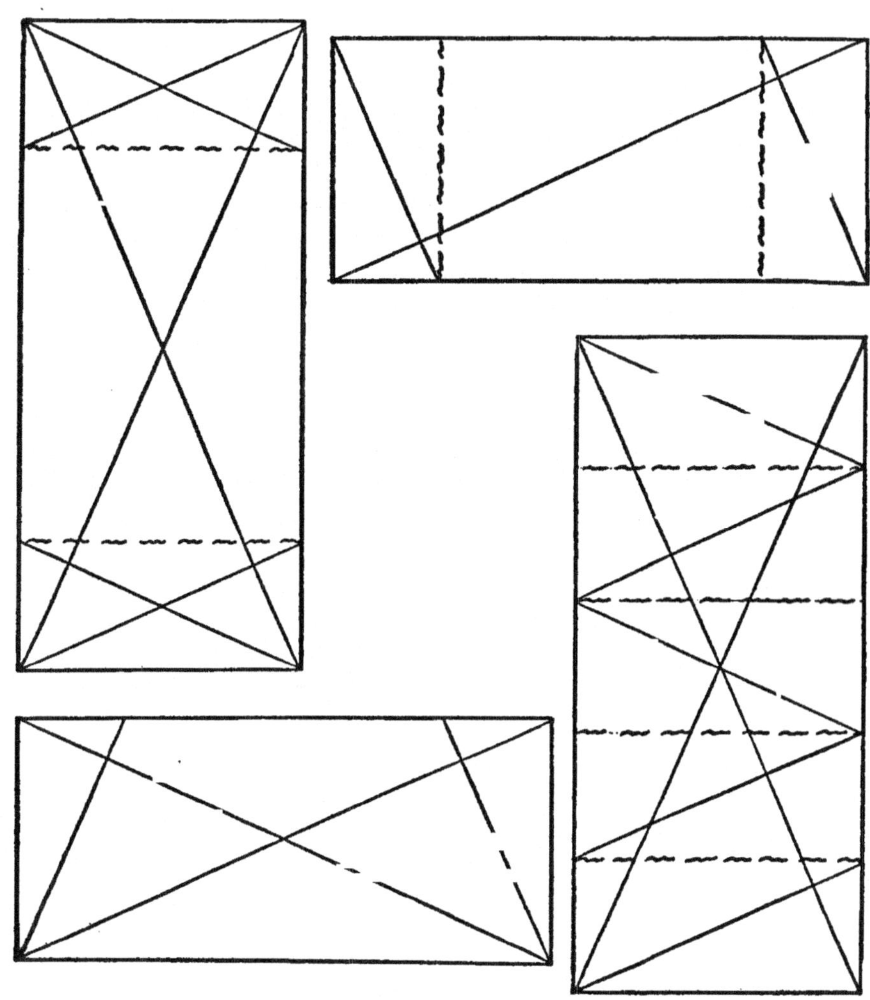

LAYOUTS IN ROOT FIVE

CHAPTER ELEVEN: COMBINED ROOTS

E HAVE already, in the preceding chapters, taken up in a simplified way the making of one root within another, using the diagonals and the squaring lines, etc., of each root to get our composition. In this chapter we will take up a few more of the combined roots in SEQUENCE, as was explained in the preceding chapters. Root One could contain two Root Four's, or a Root Two, or a Root Three, or a Five, making these inner roots by means of the quadrant arc or quarter circle and the diagonal. All of these roots could be used at one time, if so desired, inside of the major Root One.

It will be very easy to trace the roots which each grand form contains by the symbols which have been explained in Chapter Three, "Different Roots or Forms and Proportion of Pictures." (Fig. 27.)

Fig. 101 shows a layout in Root One with the diagonals with two quadrant arcs, and where these lines cross, we shall have produced two Root Two's overlapping each other, and by drawing upright lines at the same intersection, we shall have produced two Root Two's in the opposite direction overlapping each other.

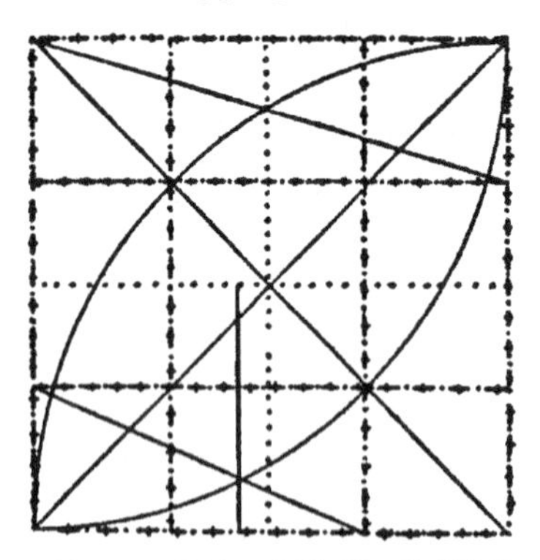

FIG. 101. ROOT ONE WITH FOUR OVERLAPPING ROOT TWO'S

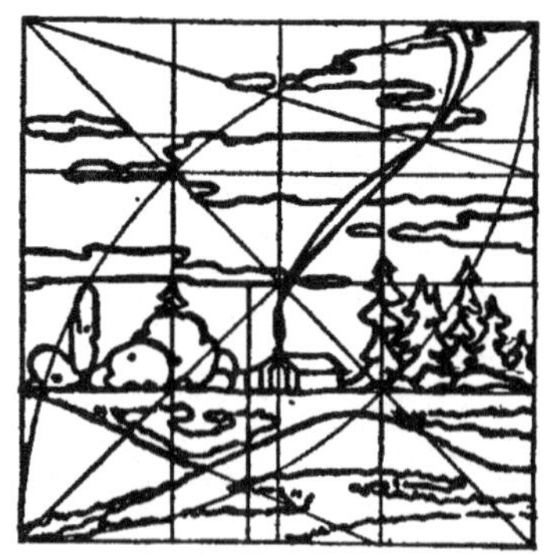

FIG. 102. PASSING CLOUDS

By drawing two crossing lines through centres we shall have made the Root One contain four Root One's also nine smaller Root One's. In Fig. 102 a composition is shown based on this layout.

In Fig. 103 a compositional layout is shown also in Root One with the diagonals; dividing this Root One in half we shall have two Root Four's. If we draw the two diagonals in each one of these Root Four's and draw parallel lines where the diagonals of the individual Root Four's cross, we shall have made a star layout, as is shown in Fig. 104, where there is a composition based on this layout.

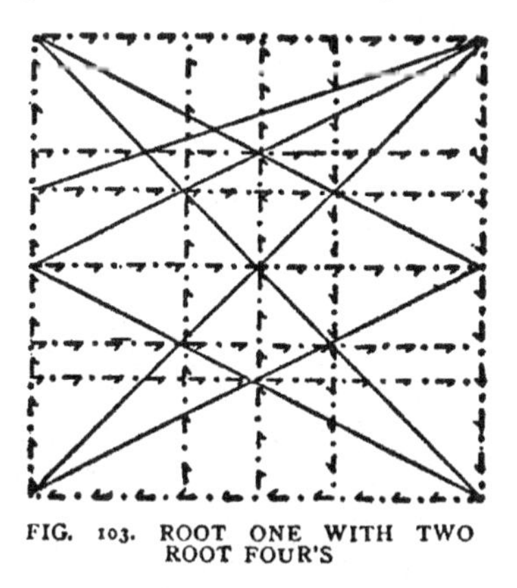

FIG. 103. ROOT ONE WITH TWO ROOT FOUR'S

FIG. 104. FIELDS

In Fig. 105 is shown a layout also in Root One, with a Root Two and a space left over. By following the symbols of the Root Two, this can be readily understood.

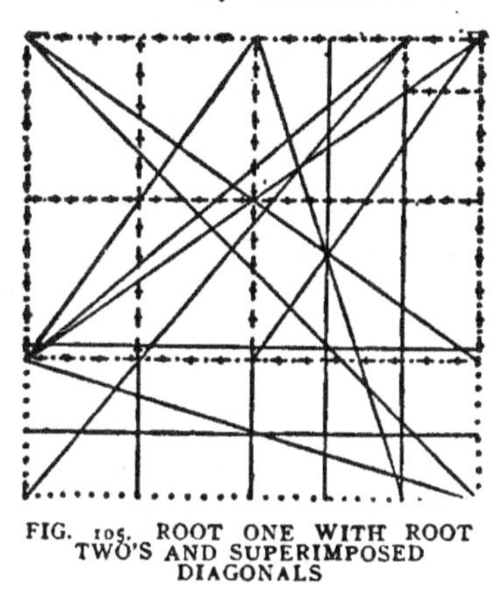

FIG. 105. ROOT ONE WITH ROOT TWO'S AND SUPERIMPOSED DIAGONALS

FIG. 106. EDGE OF THE DESERT

Taking now the Root Two so formed, and dividing it in half, we shall have made two Root Two's, because the line that crosses the diagonal meets the outer edge of the square exactly in the centre. As we know, a Root Two contains two Root Two's, and knowing this, we can draw cross lines each time through a Root Two and always produce the same form. Then, over this, we can again consider the Root One and superimpose another layout by drawing the diagonal of the Root One, and the diagonal of the remaining space from the Root Two, which was previously drawn, and all the other diagonals, as is shown; this will give us the layout, with an illustration based on this compositional form in Fig. 106.

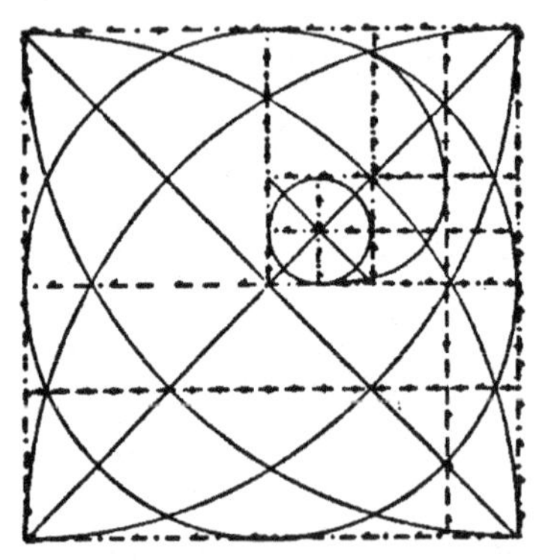

FIG. 107. ROOT ONE WITH TWO ROOT FOUR'S AND FOUR ROOT ONE'S

FIG. 108. CONVENTIONALIZED MOON

In Fig. 107, we again take the Root One, drawing four quadrant arcs and a complete circle, dividing the square in half, and one half again, giving us two squares and two Root Four's. The upper right-hand square, or Root One, we have divided so that it will contain one large Root Two and one small Root One and a small Root Two laid horizontally. It will be noted that the small Root Two divides into four Root Two's by the lines already drawn. This can be readily traced by the symbols of the dot for Root One and the spiral for the Rhythmic Curve. A composition based on this layout is shown in Fig. 108.

In Fig. 109, we have taken a Root Two as our major shape. Drawing the diagonals and the crossing lines, we shall have produced two Root Two's, side by side, and if we draw parallel lines at the intersections and diagonals at the left, where these lines bisect, we shall have made a layout, as is shown. A composition based on this layout is shown in Fig. 110.

In Fig. 111, we again take the Root Two and draw the diagonals and the crossing lines; this gives us again the so-called star layout. Then, by drawing the parallel line so as to make two Root Two's, we superimpose a layout at the intersections by making uprights where the diagonal and crossing lines meet. Then, by drawing two parallel lines at the intersections of the diagonal and the crossing line, and by drawing diagonals, we shall have made the layout, as is shown. Fig. 112 shows a composition based on this layout.

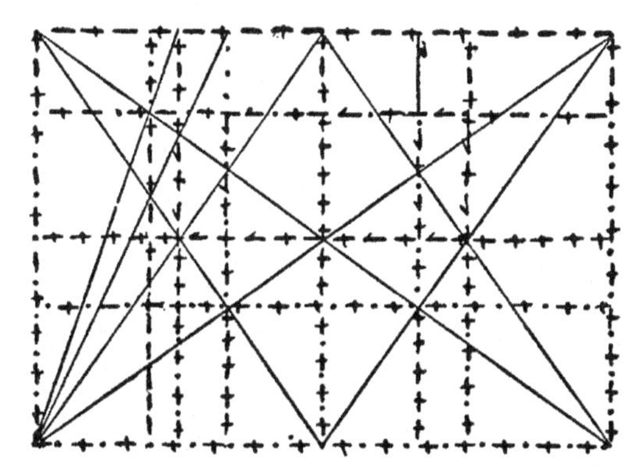

FIG. 109. ROOT TWO WITH TWO ROOT TWO'S AND PARALLEL LINES AT ALL INTERSECTIONS AND DIAGONALS AT LEFT

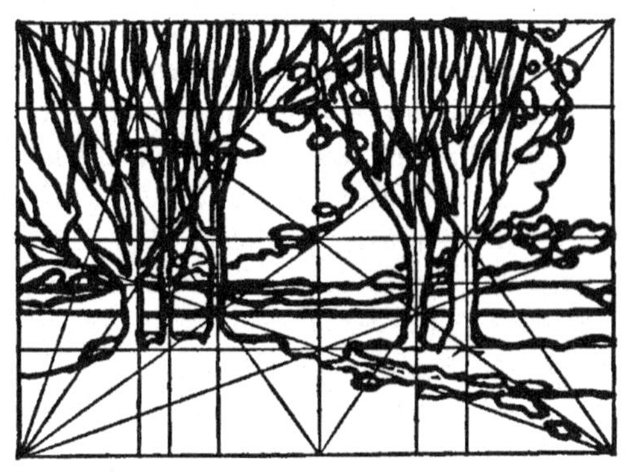

FIG. 110. WILLOWS

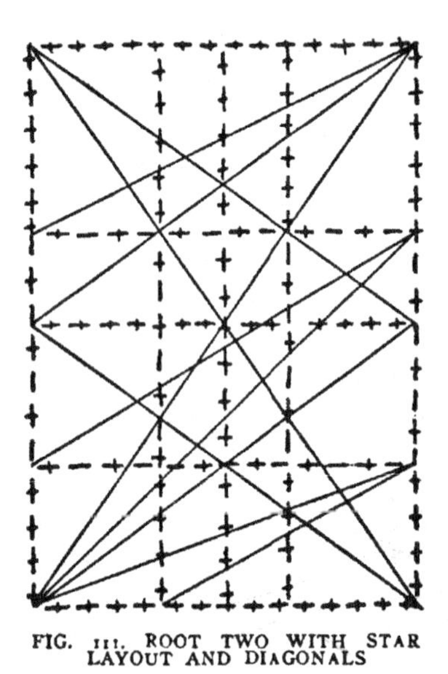
FIG. 111. ROOT TWO WITH STAR LAYOUT AND DIAGONALS

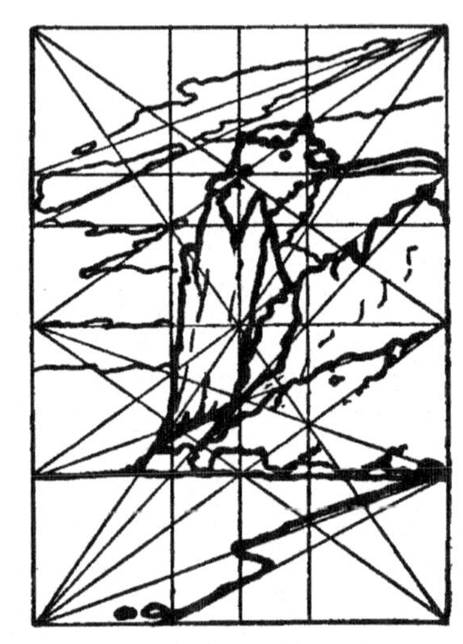
FIG. 112. BLUFFS

In Fig. 113 we have taken the Root Three and divided it by the diagonals and the crossing lines, making three Root Three's by drawing the parallel lines. In the two side Root Three's, which are in SEQUENCE, we have again drawn the diagonals. In the left-hand Root Three, we have crossed the diagonal by a series of lines at the intersections, and drawn two parallel lines through the major shape, also at the intersections. In Fig. 114 there is shown a composition based on this layout.

Fig. 115 is also a Root Three with the diagonals in the major shape, and also in the upper and lower Root Three's thus formed. Uprights have been traced through the intersections and through the centre with diagonals connecting the centre with the uprights. In Fig. 116 there is shown a composition based on this layout.

In Fig. 117, also a Root Three, by drawing again the three Root Three's in SEQUENCE in the same manner as before explained, and drawing the diagonals and parallels at intersections, you will have completed the layout. In Fig. 118 is shown a composition based on this layout.

Fig. 119 is also a Root Three with six Root Three's and diagonals to the half for a newspaper or magazine layout. The various shapes shown,

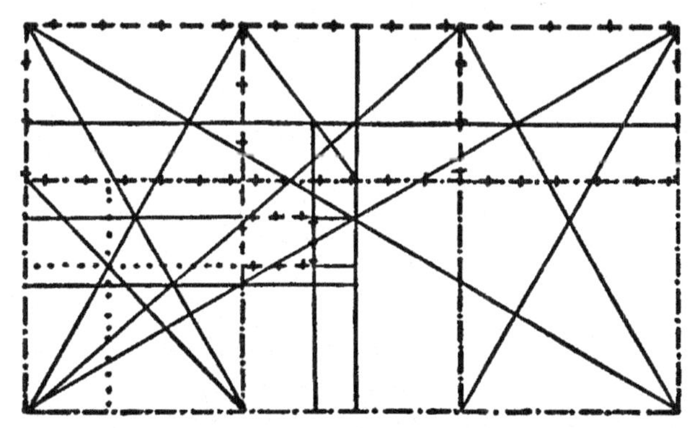

FIG. 113. ROOT THREE WITH THREE ROOT THREE'S WITH DIAGONALS AND PARALLEL LINES

FIG. 114. PANDORA

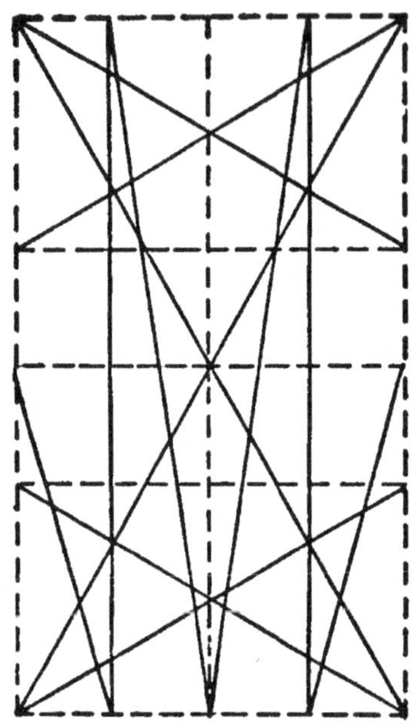

FIG. 115. ROOT THREE WITH THREE ROOT THREE'S UPRIGHT, PARALLELS AND DIAGONALS

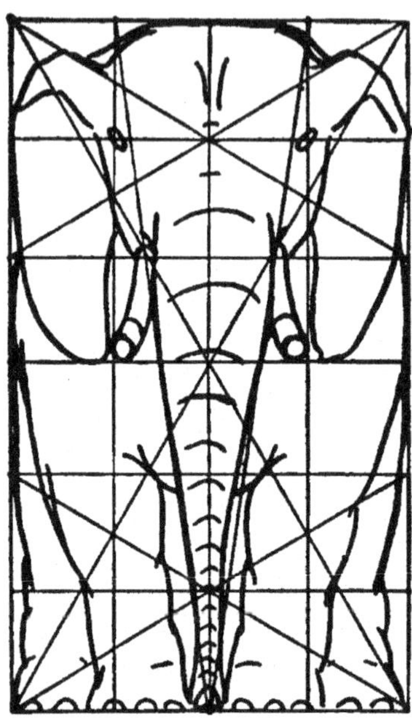

FIG. 116. CONVENTIONALIZED ELEPHANT

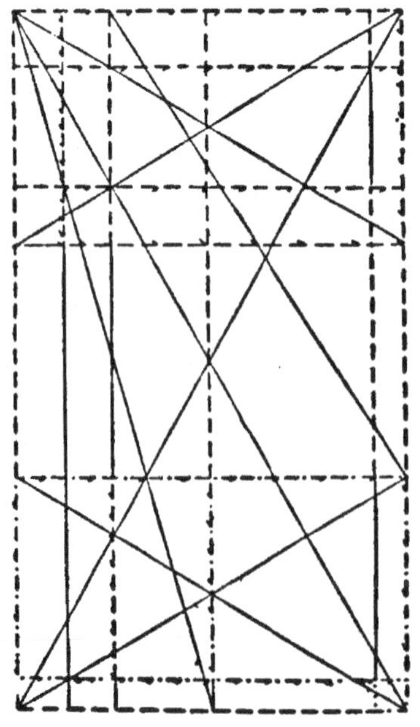

FIG. 117. ROOT THREE WITH THREE ROOT THREE'S AND NUMBERS OF PARALLELS AT INTERSECTIONS

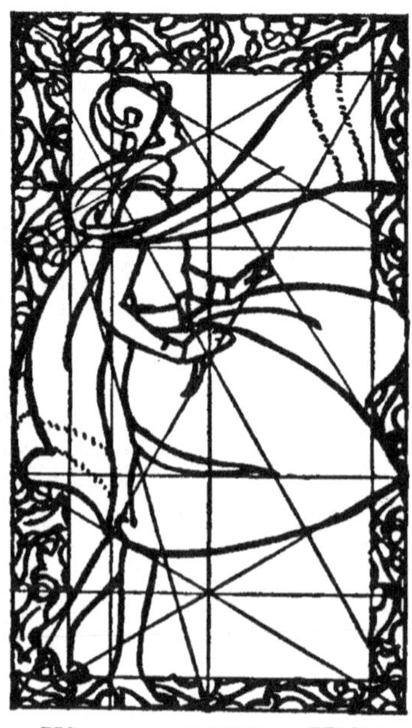

FIG. 118. A BORDER PATTERN

of course, have each a relation to the other, on account of using the Dynamic lines. This must always be true. In Fig. 120 is shown a composition based on this layout.

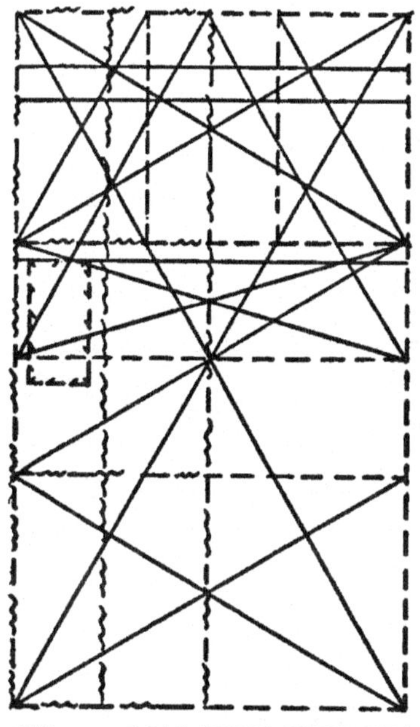

FIG. 119. ROOT THREE WITH SIX ROOT THREE'S AND DIAGONALS TO THE HALF

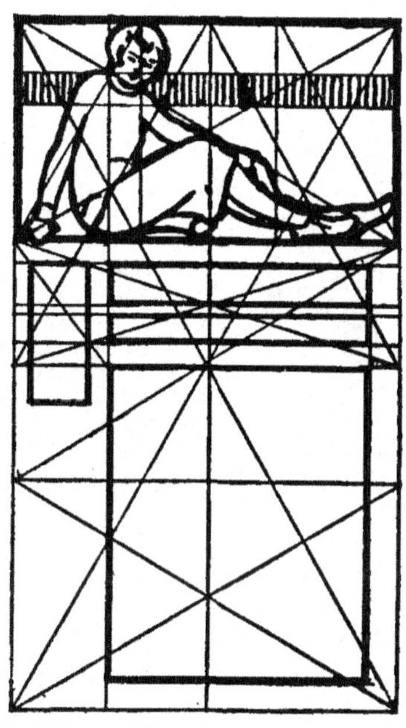

FIG. 120. COMMERCIAL LAYOUT

Fig. 121 is a Root Four which shows the major shape divided into four Root Four's by drawing the two diagonals with the crossing line and the parallel line, also a parallel line through the centre; by drawing upright lines through the intersections with the diagonals you will have completed this layout. In Fig. 122 is shown a composition based on this layout, which shows that it is not always necessary to follow straight lines, but the line can be curved in the form of the Whirling Square to make a rhythmic composition.

In Fig. 123 is shown a Root Four with the spiral lines; by drawing the diagonal and the crossing of the diagonal, you will have produced a Root Four on each end and a Root One in the centre. By drawing a parallel at

PEONIES AFTER AN OIL PAINTING BY MICHEL JACOBS

In overlapping Root Ones

ROCK OF ALL NATIONS — By MICHEL JACOBS

Modeled in bronze in Root Four

the intersection of the diagonals, and by cutting the lines up with the spiral, you will have completed the layout with rhythmic lines, but not in the dimensions of the Whirling Square which has been carried out in a composition in Fig. 124.

In Fig. 125, you have a Root Five divided up into two Root Five's, one on each end, by means of the diagonal and the crossing line, and the rhythmic line drawn from the intersection in the left-hand upper corner. This layout has been carried out in composition in Fig. 126.

Fig. 127 shows the Whirling Square Root with a composition based on this layout. (Fig. 128.) This last composition shows the possibility of combining the rhythmic, flowing line with Dynamic Symmetry.

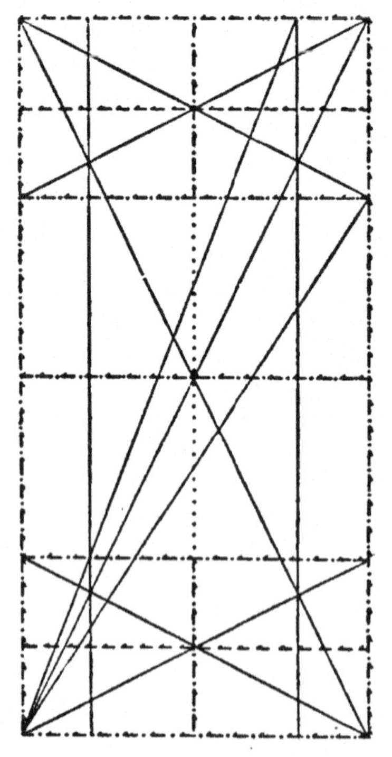

FIG. 121. ROOT FOUR WITH FOUR ROOT FOUR'S AND PARALLEL LINES THROUGH INTERSECTIONS AND DIAGONALS

FIG. 122. COMPOSITION BASED ON FIG. 121

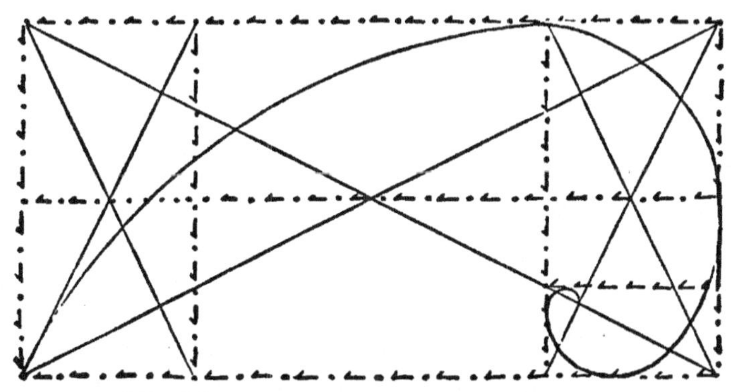

FIG. 123. ROOT FOUR WITH TWO ROOT FOUR'S AND ONE ROOT ONE USING THE RHYTHMIC LINES

FIG. 124. THE WAVE

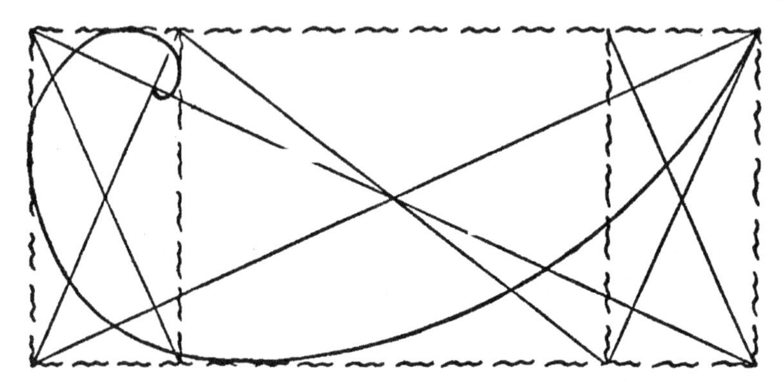

FIG. 125. ROOT FIVE WITH TWO ROOT FIVE'S AND RHYTHMIC LINES

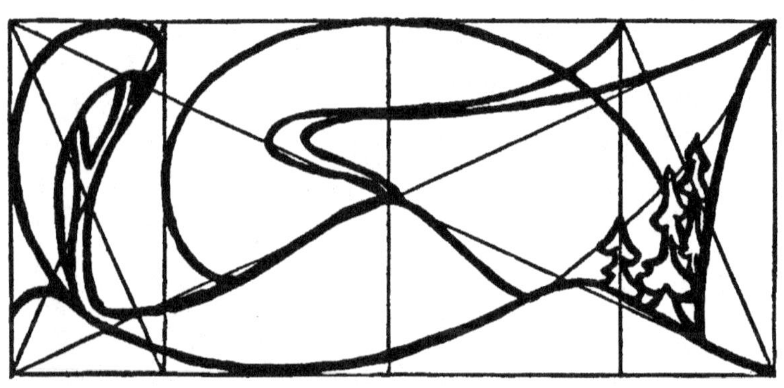

FIG. 126. THE SLOPE

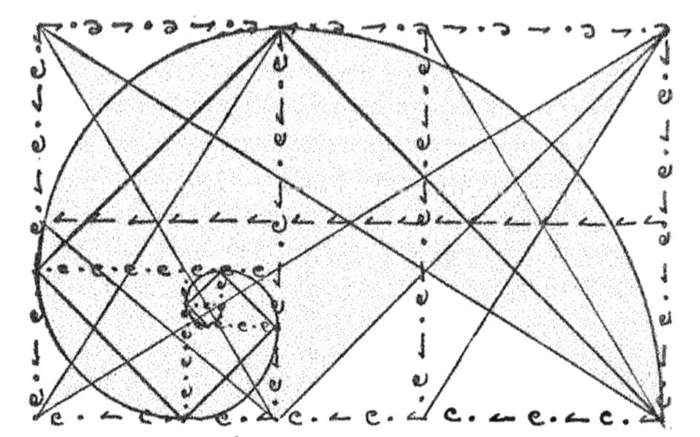

FIG. 127. WHIRLING SQUARE ROOT AND DIAGONALS

FIG. 128. CONVENTIONAL PATTERN

CHAPTER TWELVE: MORE COMPLEX COMPOSITIONS

E HAVE seen in the previous chapter a method whereby we used more than one root in the major shape. In this chapter, we will take up both two and three roots in the major shape. It will not be necessary to describe in detail why we have drawn each of the lines in the individual illustrations, as I believe, now, that the reader is conversant with the general scheme, but I wish to call to the attention of the student of Dynamic Symmetry a few cardinal points.

Any major shape may contain other roots which may be found by drawing the diagonal, squaring the diagonal, and by drawing the parallels where this crossing line meets the outside of the large form; and also the major shapes may be divided in many ways, as is shown on page 100.

Root One:

It can contain all the other roots by means of drawing the quadrant arc or quarter circle, drawing the diagonal and the parallel line at intersections leading from Root Two down to Root Five.

The Root One also can be divided into four equal parts making four Root One's, so that one root can overlap another root. For example, you can get two Root Two's in a square by having one overlap the other.

Also, this Root One can be made to contain two Root Four's by dividing the square in half so that each one of these Root Four's would contain two Root One's.

Each one of these smaller forms or forms in Sequence, as we know them, can be divided with the diagonal, the squaring line, and the parallel line.

Root Two:

It can contain two Root Two's by simply dividing the length in half, or by drawing the diagonal, the crossing line, and the parallel line.

The Root Two can also be divided into three equal parts by making the parallel line pass through the intersection of the crossing line and the diagonal, and by making the rhythmic curve, as is shown in Fig. 129. In Fig. 130 is a composition based on this layout.

The Root Two can also be divided into four Whirling Square Roots, one overlapping the other, by means of making the square on one end of Root Two, as in Fig. 131, dividing this square into four smaller

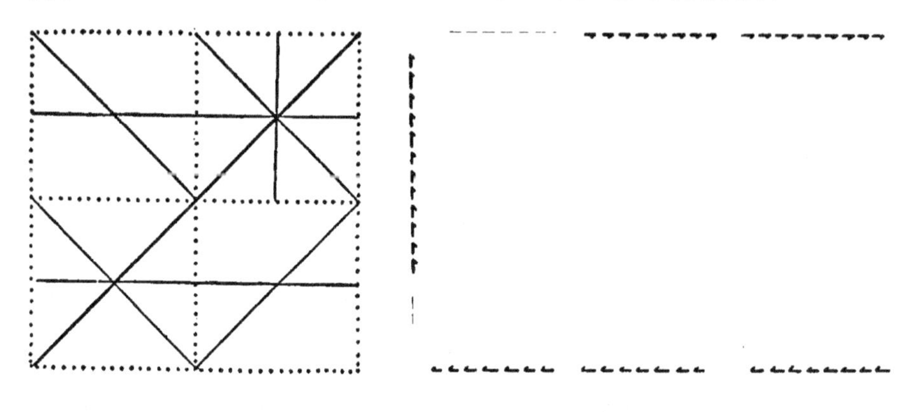

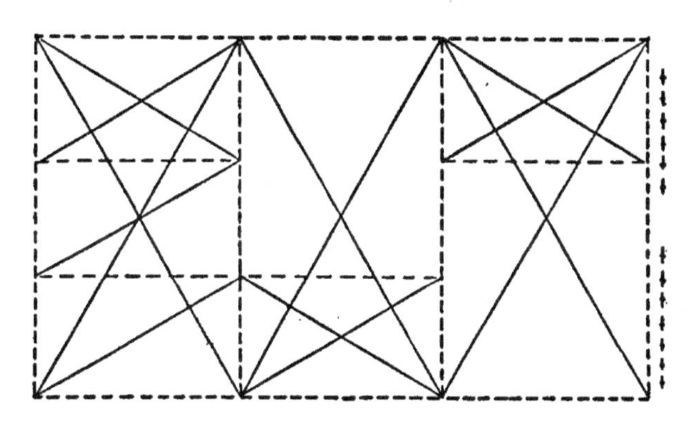

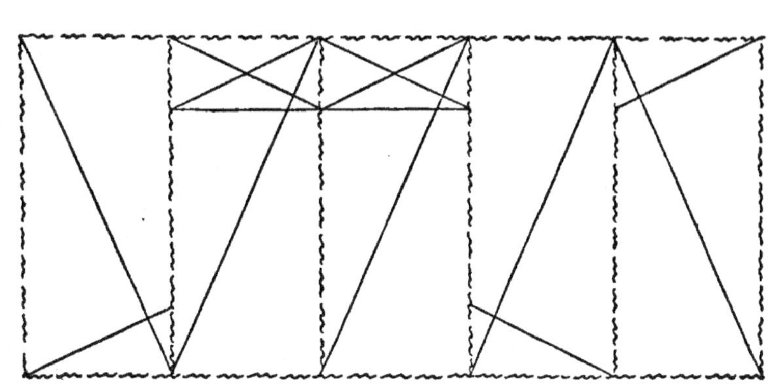

MAJOR SHAPES DIVIDED INTO COMPLEX FORMS

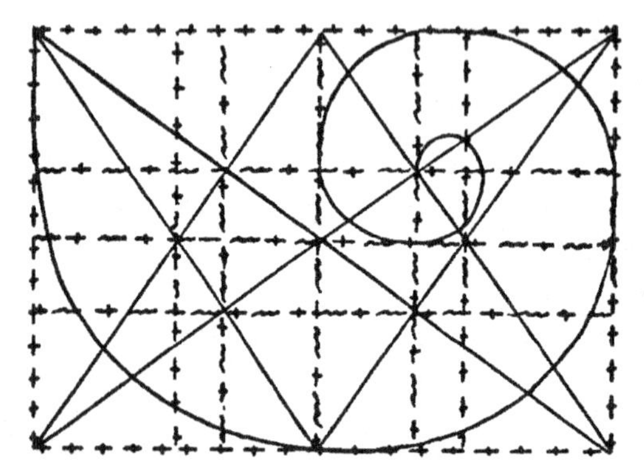

FIG. 129. ROOT TWO DIVIDED INTO THREE EQUAL PARTS AND USING THE RHYTHMIC CURVE

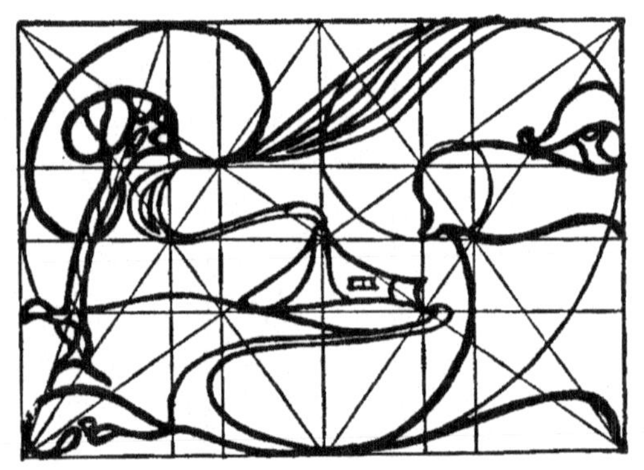

FIG. 130. CONVENTIONAL LANDSCAPE

squares, taking the half of one of these small squares, finding out the diagonal of the half, and making a Whirling Square Root in one corner, as in Figs. 132 and 133, and repeating it in the other three corners, as in Fig. 134.

A few of the many subdivisions or forms in SEQUENCE will be found on page 70.

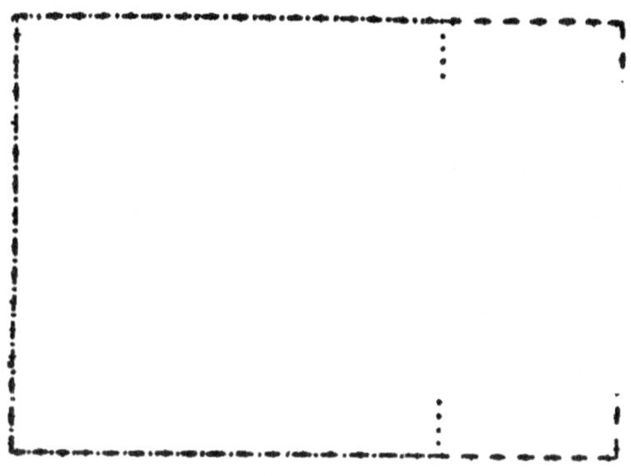

FIG. 131. ROOT TWO WITH A ROOT ONE ON SIDE

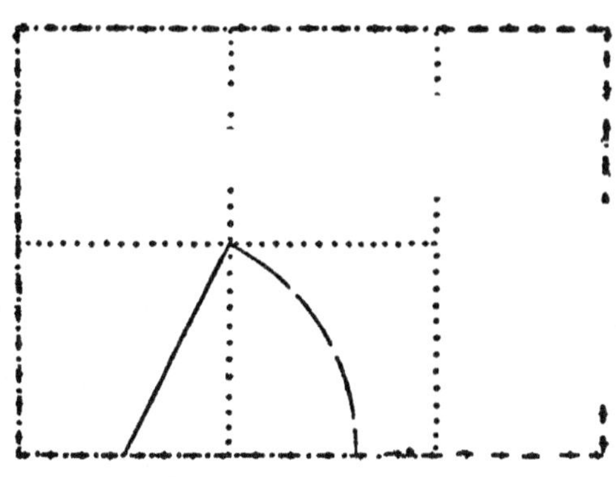

FIG. 132. ROOT TWO SHOWING HOW TO MAKE THE WHIRLING SQUARE

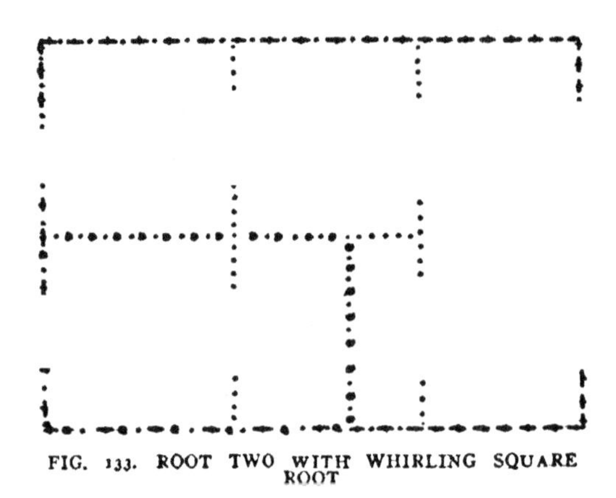

FIG. 133. ROOT TWO WITH WHIRLING SQUARE ROOT

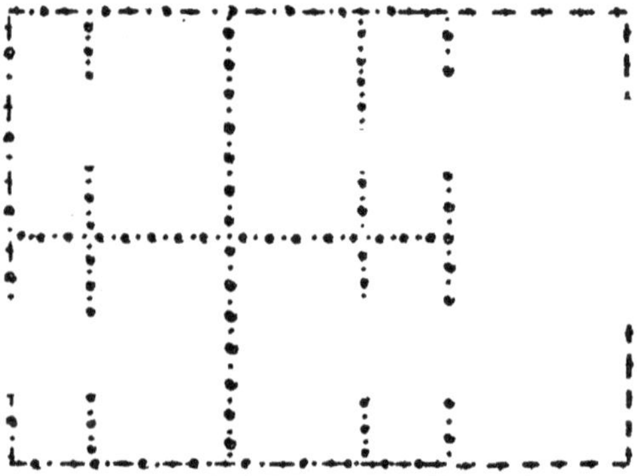

FIG. 134. ROOT TWO WITH FOUR WHIRLING SQUARES

Root Three:

It can be divided into three equal divisions, as is shown on page 71. Numbers of different forms can be drawn into this Root Three, and many have been shown on page 74.

Root Four:

It can be divided into two Root One's or four Root Four's, as is shown on page 78. Each one of the Root One's can be divided in their turn by all the subdivisions as shown on page 55 under Root One. On page 81, many layouts are shown within Root Four.

Root Five:

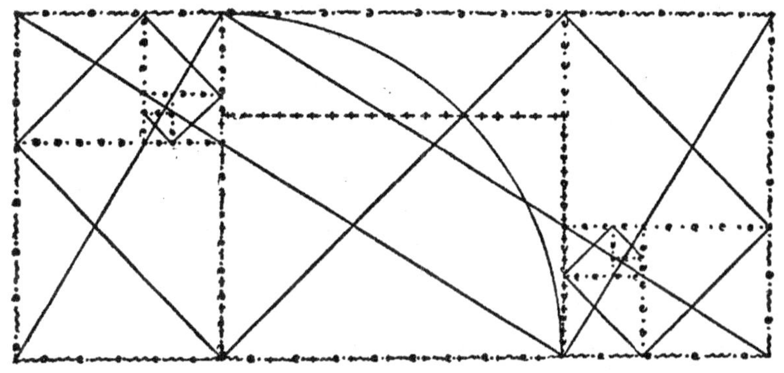

FIG. 135. ROOT FIVE WITH TWO WHIRLING SQUARES OVERLAPPING AND A WHIRLING SQUARE ON EACH END WITH ROOT TWO INSIDE OF A ROOT ONE

FIG. 136. PETER RABBIT

It is a very important root. It can be divided into five equal parts; each one of the lesser forms will be Root Five's.

It can be divided into one square and two Whirling Square Roots, as is shown on page 29. Each one of these forms in SEQUENCE can be laid out or subdivided, as is shown under their respective headings in this chapter.

We could take each one of these Whirling Square Roots and form the Whirling Square Root in SEQUENCE, as is shown in Fig. 135. By looking at this illustration, you will see that we have combined in the Root Five two Whirling Square Roots, one Root One and one Root Two. In Fig. 136 there is shown an illustration based on this layout.

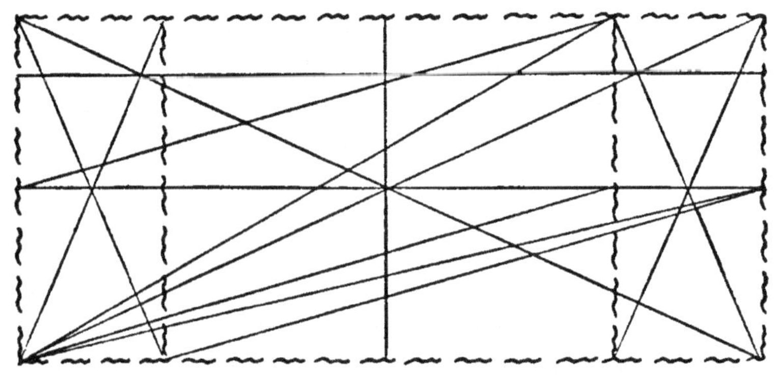

FIG. 137. ROOT FIVE WITH A ROOT FIVE AT EACH END WITH DIAGONALS AND PARALLEL LINES

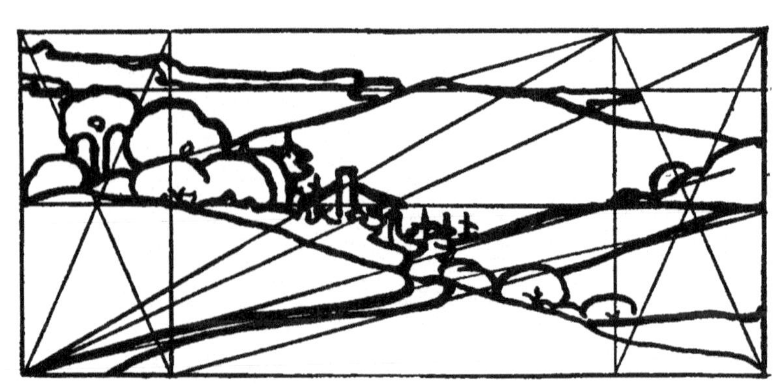

FIG. 138. ROLLING GROUND

In Fig. 137 is shown a layout in Root Five by means of the diagonal, the crossing line, the parallel line, and numbers of diagonals in the forms so constructed; and in Fig. 138 is shown an illustration based on this layout.

In Fig. 139 is shown another composition in Root Five, which is also made with the diagonal, crossing line, parallel lines, and numbers of diagonals differently arranged from those in the preceding layout. In Fig. 140 there is shown a composition based on this layout.

Whirling Square Root:

It is extremely interesting on account of its association with Root Five,

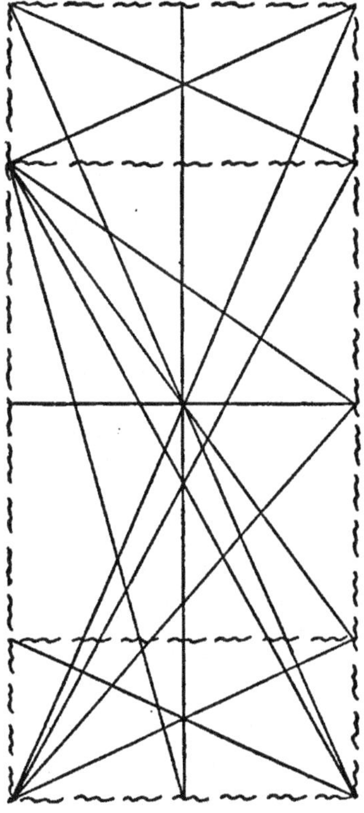

FIG. 139. ROOT FIVE WITH A ROOT FIVE ON EACH END WITH DIAGONALS FROM CORNERS AND PARALLELS THROUGH CENTRE BOTH WAYS

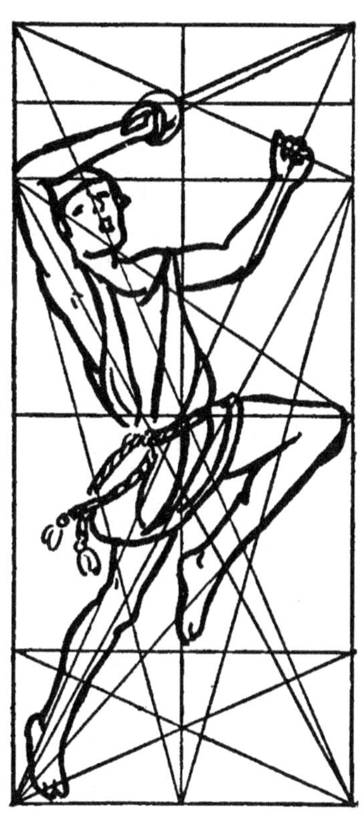

FIG. 140. WARRIOR

which is the basis of all Dynamic Symmetry. As I have shown under Root Five, two Whirling Square Roots and a square make a Root Five. I have shown on page 116 a number of subdivisions and arrangements of the Whirling Square Root. Any one of these subdivisions can, of course, have the diagonals, the crossing line, and the parallel line. On pages 114 and 115 I have shown a number of different layouts with illustrations based on them which can be readily understood at this time.

In Fig. 141 there is shown a shape which is less than a Root Two, to demonstrate that it is possible to use, if necessary, a form which does not fit any of the roots. In Fig. 142 there is shown an illustration based on this layout.

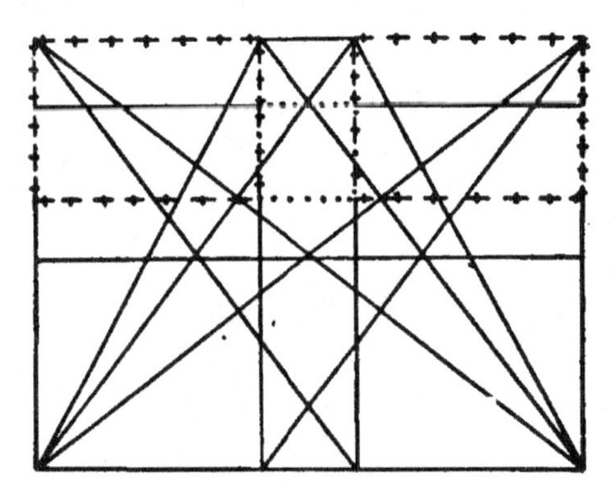

FIG. 141. A FORM LESS THAN ROOT TWO WITH FORMS OVERLAPPING

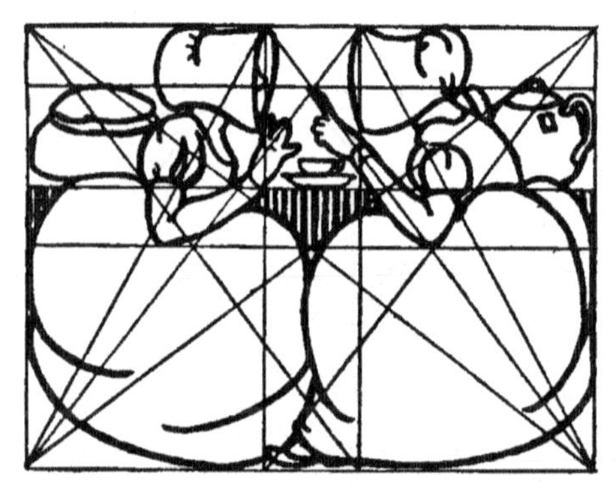

FIG. 142. THE GOSSIPS

I trust that the reader, at this time, understands sufficiently the principals of Dynamic Symmetry to know that it is possible to make innumerable lay-outs and combinations of forms to suit the conception of the artist, and in closing this chapter, I wish to emphasize the fact that it is always necessary first to visualize the conception. Even go so far as to draw the picture, or, at any rate, the sketch, with your original conception fresh in your mind. Then, finding out which of the roots and Dynamic lines will more nearly carry out your idea, change the line so that it will come nearer to the Dynamic line.

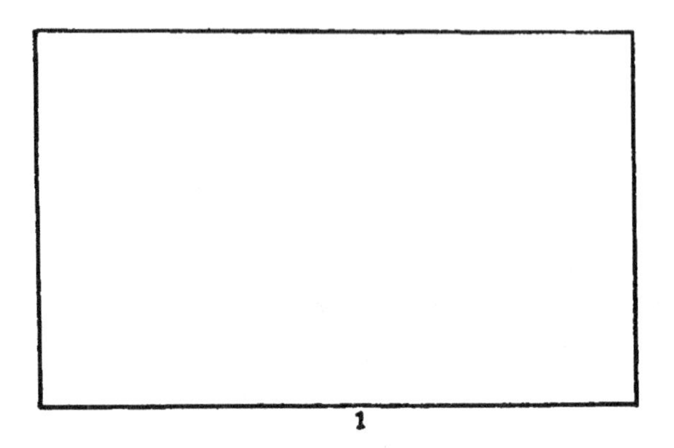

1

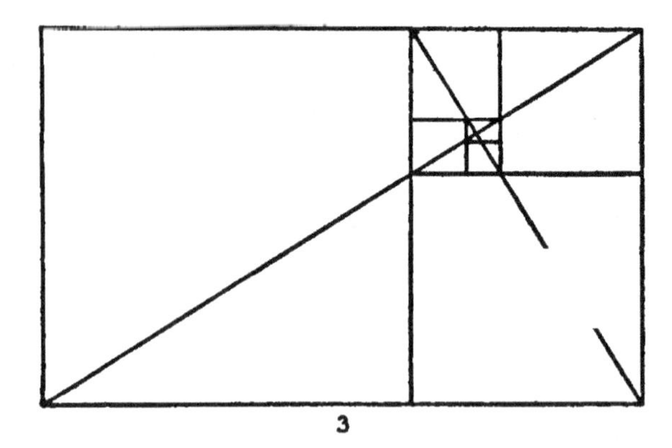

3

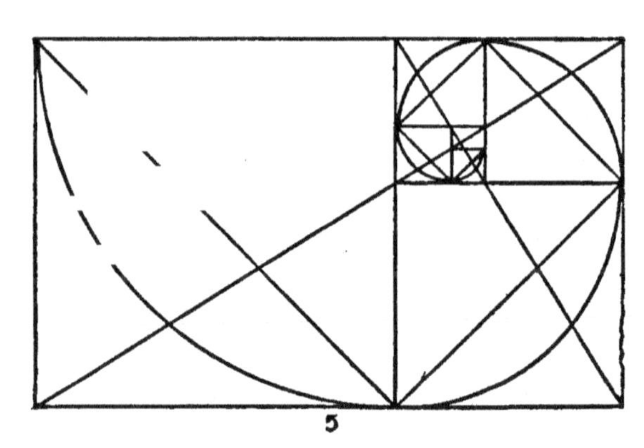

4 5

PROGRESSIVE STEPS OF THE WHIRLING SQUARE ROOT

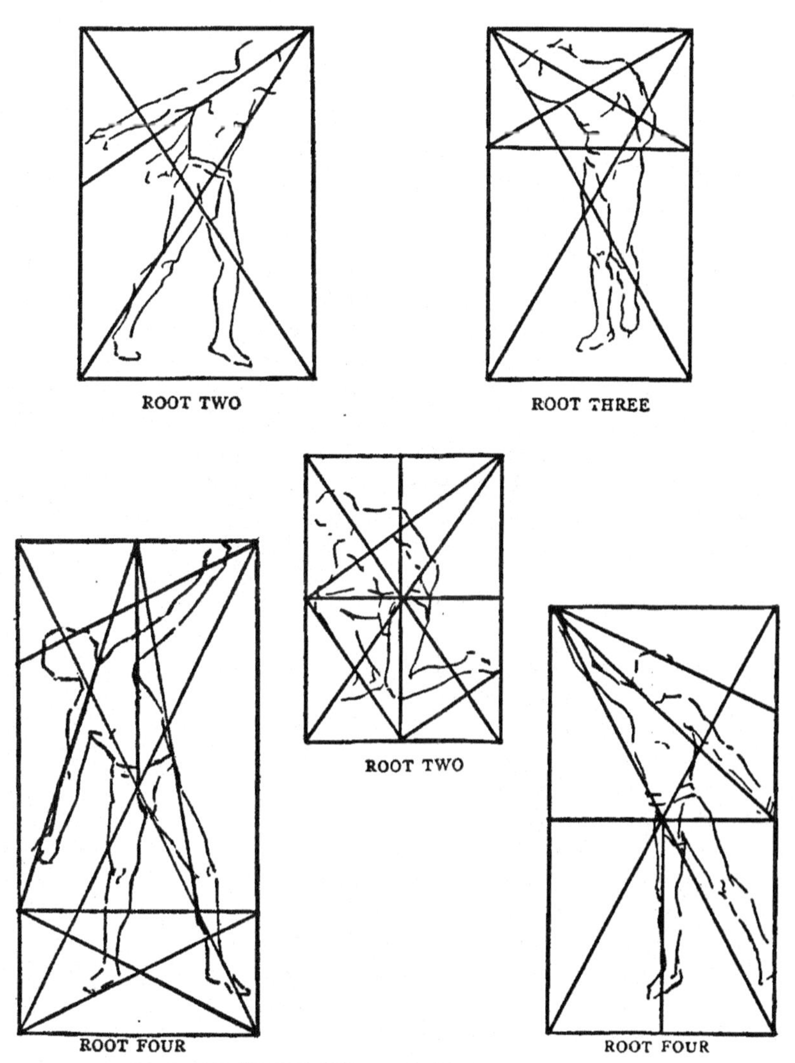

DIAGRAMS OF DYNAMIC POSES OF THE HUMAN FIGURE

DYNAMIC POSES OF THE HUMAN FIGURE

DYNAMIC POSES OF THE HUMAN FIGURE

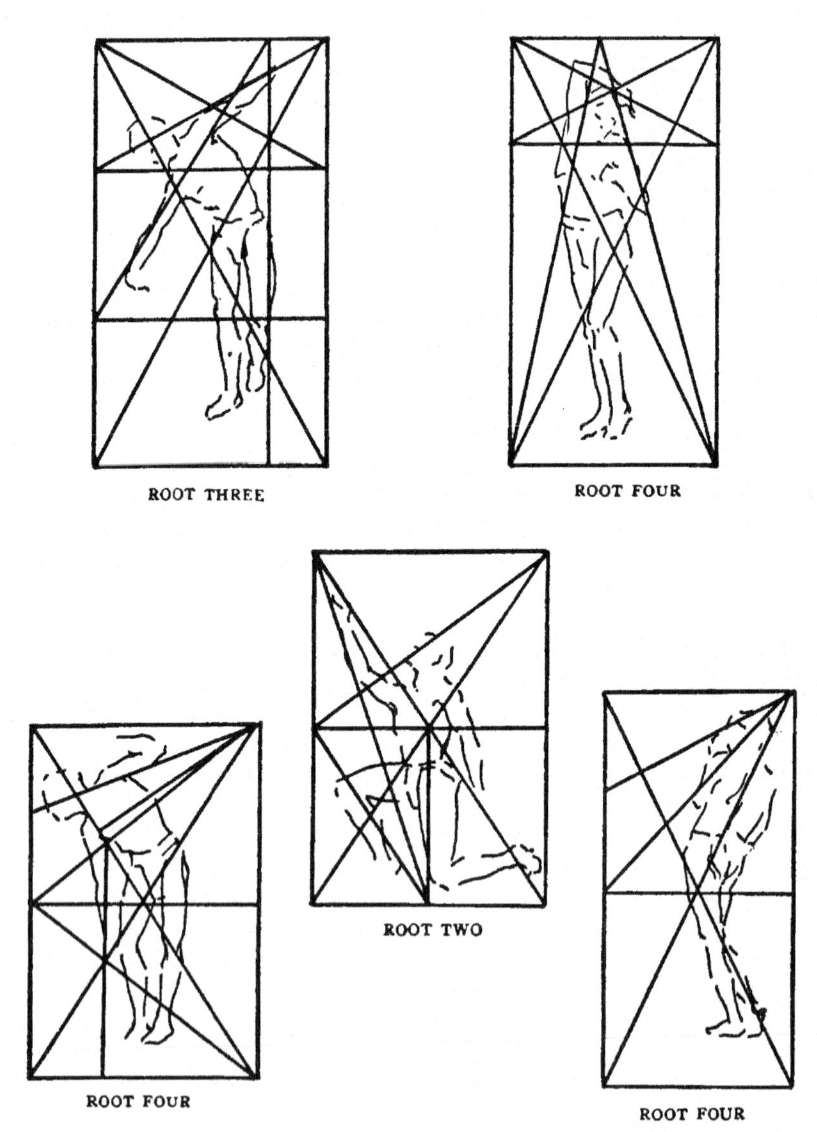

DIAGRAMS OF DYNAMIC POSES OF THE HUMAN FIGURE

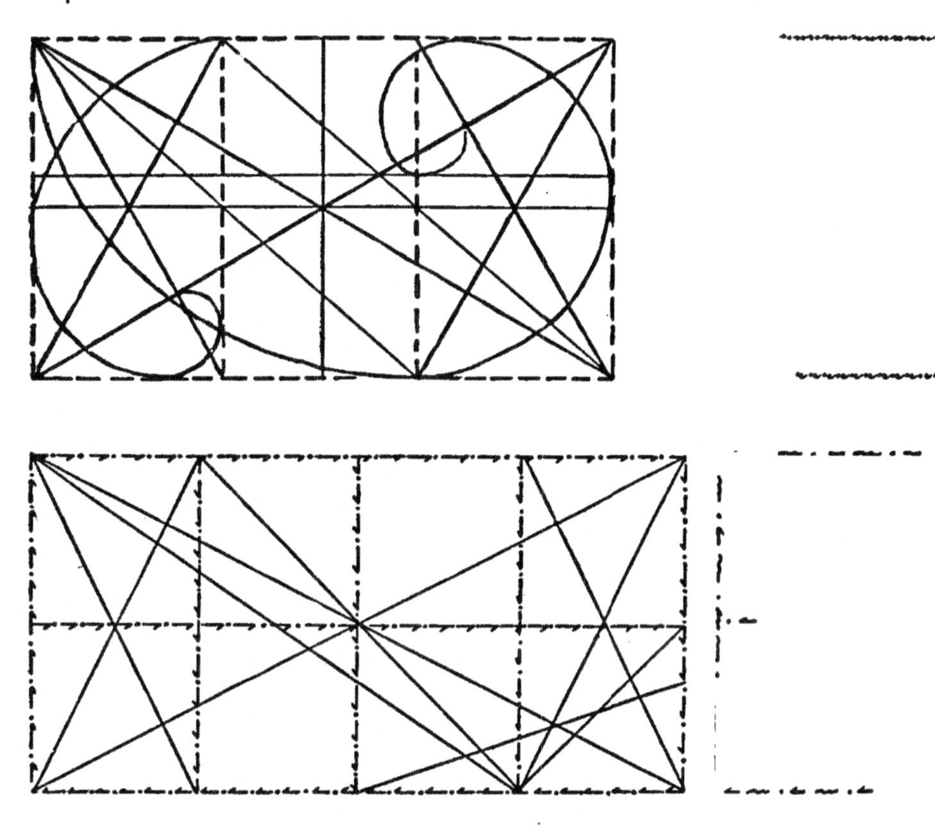

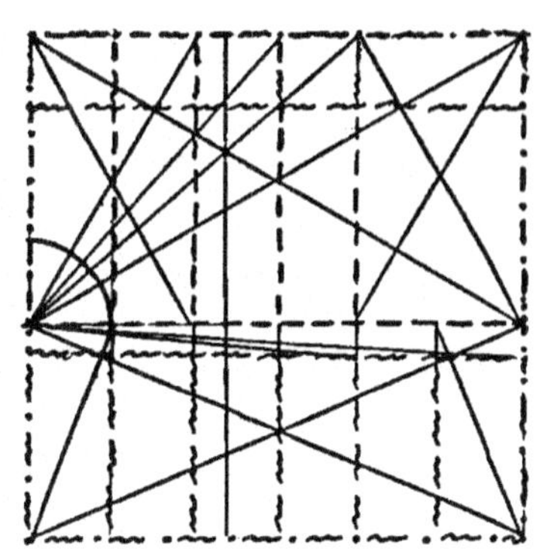

LAYOUTS OF COMPLEX COMPOSITIONS

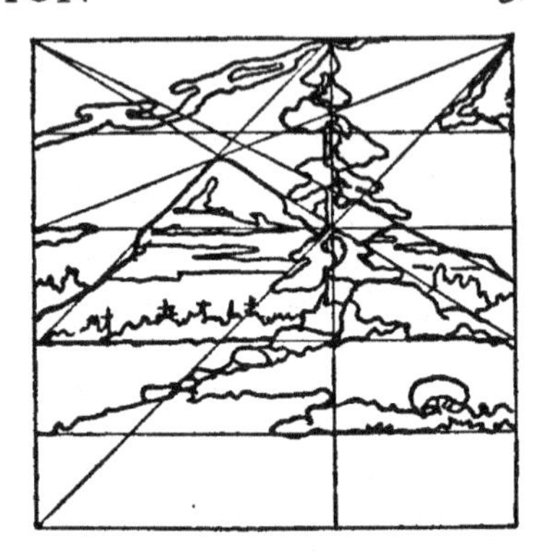

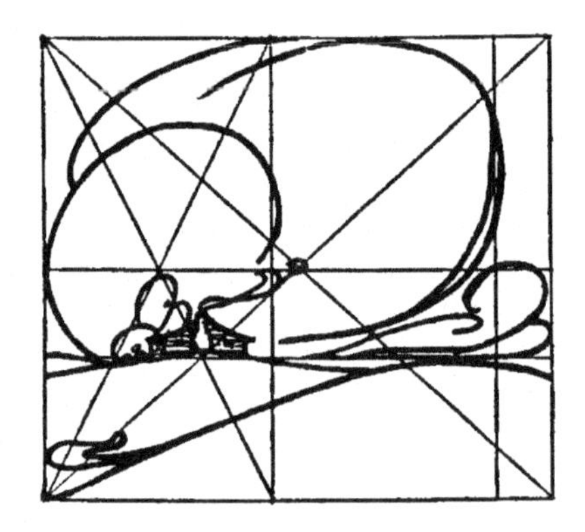

ILLUSTRATIONS OF COMPLEX LAYOUTS

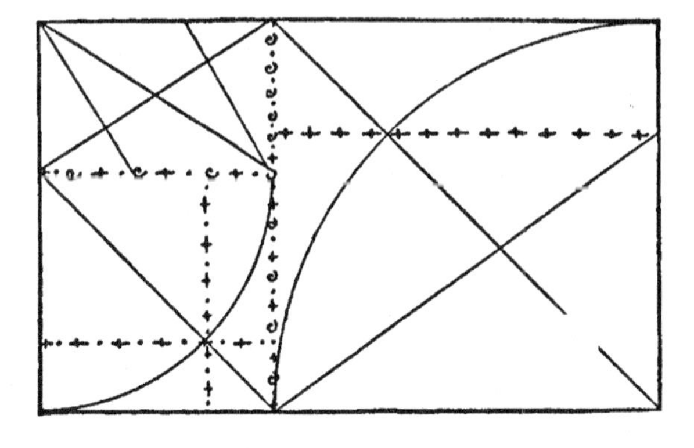

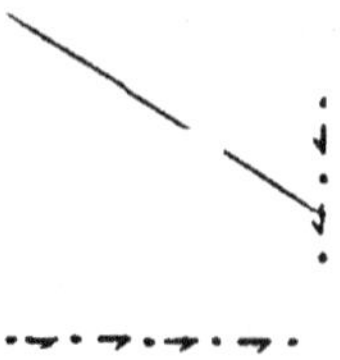

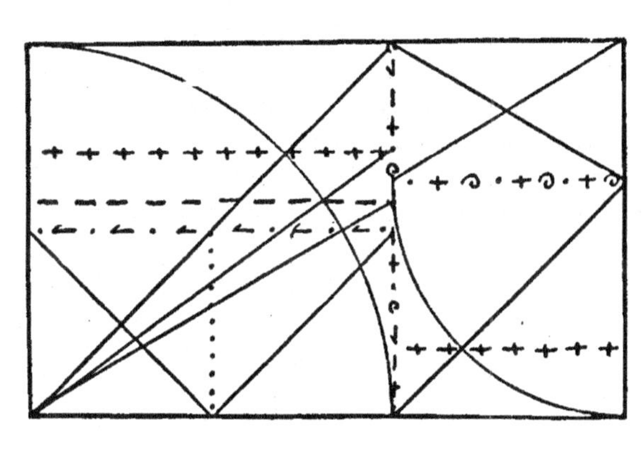

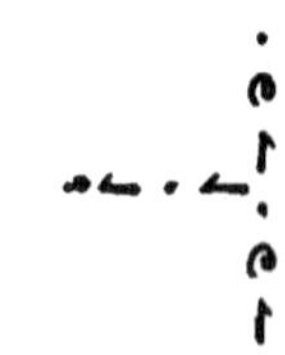

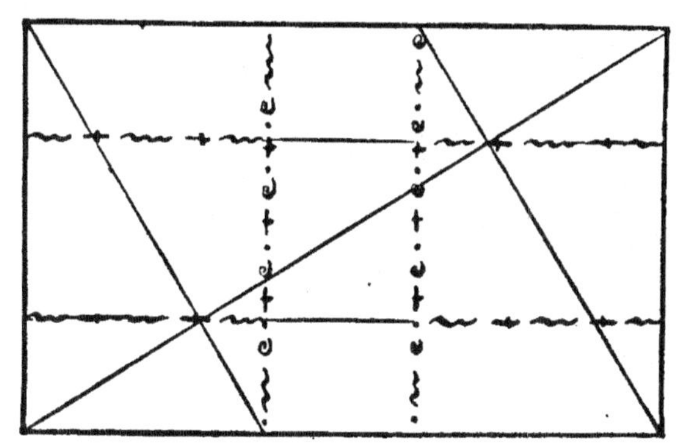

LAYOUTS IN THE WHIRLING SQUARE ROOT

CHAPTER THIRTEEN: GROUND COMPOSITION IN PERSPECTIVE SHOWING THE THIRD DIMENSION

E HAVE seen in the preceding chapters the composition or arrangement of masses so as to make a decoration in the two dimensions. Of course, the idea of design or pattern which all modern pictures should have is very important, whether they are painted in the subjective, which is just the spirit of the thing, or objective, which is a more or less literal translation of realistic forms. The picture must have a pattern, in the same way as a piece of cloth or other material.

This is especially true in the subjective form, and is more noticeable in the two dimensions, but it is also necessary to have a pattern in the ground, in the third dimension. This is very obvious in sculptural or architectural work, but it is also necessary on the canvas or paper.

The better way to understand this problem would be to draw a parallel perspective, or, as it is sometimes known, one point perspective, and in this perspective drawing lay out the root, and inside this root lay off your diagonals and your crossing lines. In Fig. 143 there is shown a perspective drawing with a Root One or square laid out with the diagonal and the quadrant arc in perspective with a parallel line which shows the Root Two.

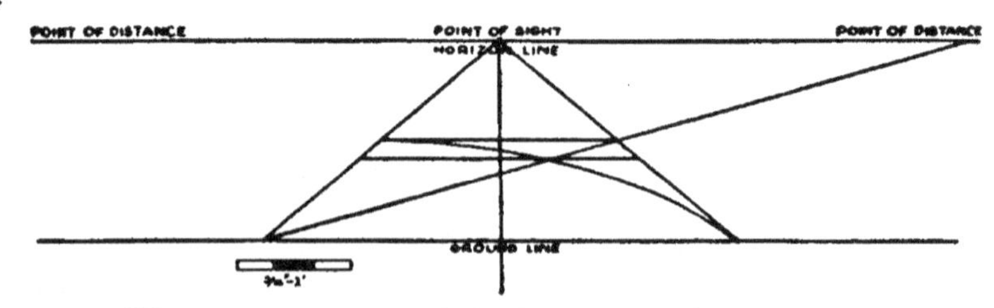

FIG. 143. PERSPECTIVE OF ROOT ONE WITH A ROOT TWO

I have drawn this illustration in scale the better to convey the idea in mind. But, of course, the artist will not use any mathematical measurements for his picture, but will approximate the placement of these lines so as to arrange his ground composition in the three dimensions to the best advantage. Fig. 144 is a composition based on this layout. In Fig. 145 there is shown a Root Five in perspective divided by two Whirling Square Roots with the spiral or Whirling Square designated. In Fig. 146 there is carried out a composition based on this layout.

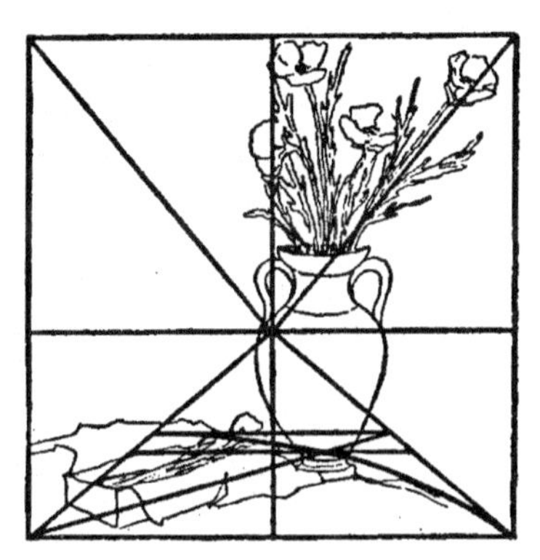

FIG. 144. COMPOSITION BASED ON FIG. 143

If the reader wishes to go to the trouble of making a layout on transparent paper of the root that he wishes to use in two dimensions, holding this up in front of the perspective layout in the third dimension, he will have one overlapping the other.

I suggested in Chapter Four that the artist make different roots with different layouts on transparent guides with waterproof ink. If he holds these up to the eye (as was explained in that chapter), standing off from the picture, he will be able to judge immediately the corrections to be made in the original sketch or painting, making it conform nearer to the root he wishes to use.

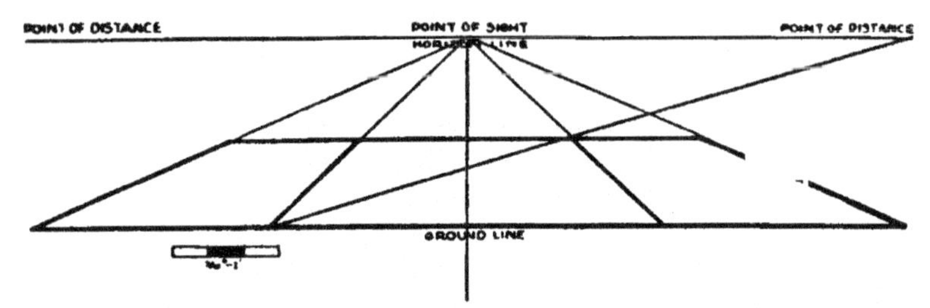

FIG. 145. PERSPECTIVE OF ROOT FIVE WITH TWO WHIRLING SQUARE ROOTS

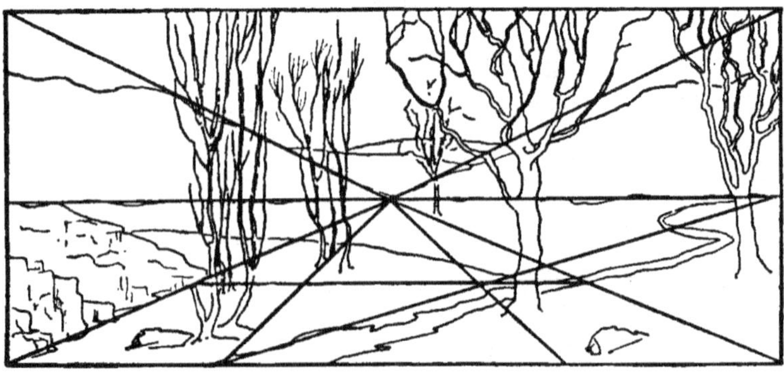

FIG. 146. COMPOSITION WITH PERSPECTIVE GROUND

CHAPTER FOURTEEN: COMPOSITION OF MASS, LIGHT, AND SHADE

AS WE explained in the first part of this book and illustrated with the seesaw, the weight or mass of any composition must be considered. While Dynamic Symmetry will give you the placements of the principal points and other points of interest in SEQUENCE, and will give you, to a certain extent, the shape of the masses to be followed (as closely as possible to the original conception of the artist), it must always be remembered that the weight of these masses would throw your picture off balance if they were not considered.

For example: if we were to take a layout and paint the principal point of interest a gray, and another part, which we intended to keep as a minor point of interest, a black surrounded by a white mass, the principal point of interest would not hold our attention.

It must always be borne in mind that the greatest contrast in black-and-white value will attract the eye. Sometimes we put a very light highlight into a dark mass, and sometimes the reverse—putting a dark mass into a light area: either one of these methods will hold the eye.

FIG. 147. SUPPLICATION TO ZEUS. (DARK MASS BELOW AND LIGHT MASS ABOVE)

FIG. 148. WAR. (DARK MASS ABOVE AND LIGHT MASS BELOW

Then, again, the texture will either attract or will not hold your attention. For example: a black-and-white stripe of heavy lines wide apart will hold the eye, and a dotted line will give atmosphere. Oblique lines will not attract as much attention, or hold up in the foreground as much as the perpendicular or horizontal. On page 125 I have shown you a few textures reproduced by means of the Ben Day process, to remind you of these textures in the composition.

Another example of balance or weight of composition, as we explained with the seesaw, is that one large mass can equal two or more smaller masses. (See Fig. 9.)

If the large mass of dark colour is used in the composition at the lower part or near the ground, and the light mass in the upper parts, it will give a feeling of rest and solidity. (Fig. 147.) If the reverse of this is done, and the dark mass is in the upper part and the light mass below, it will give a feeling of overbalance or action, which is sometimes desired. (See Fig. 148.)

Another thing that must be borne in mind is that a very dark mass will seldom balance a very light mass, except, as we have said before, that the dark mass is in the lower parts of the picture, as is shown in Fig. 147.

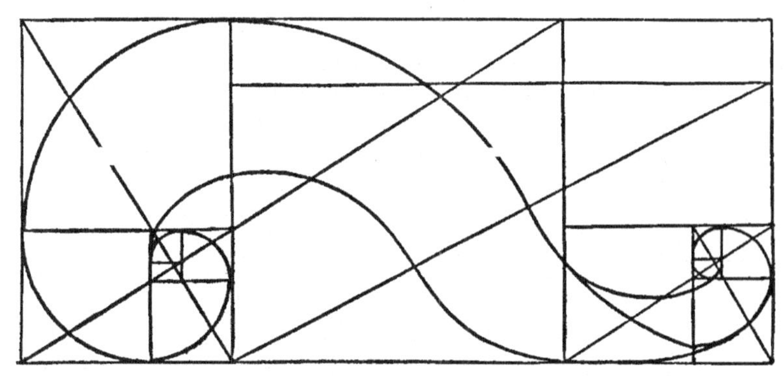

FIG. 149. WHIRLING SQUARE TO SHOW MASS, LIGHT AND SHADE

Combining Dynamic Symmetry with light and shade, one will see in Figs. 149 and 150 two roots of the Whirling Square. Both of these have been carried in SEQUENCE to the smaller Whirling Square and darkened to show the effect of the graduation of dark and light.

FIG. 150. WHIRLING SQUARE TO SHOW MASS, LIGHT, AND SHADE

This leads the eye from the larger spiral in mass to the smaller form, and vice versa.

Lines which surround masses, if very dark, will appear to be part of the background, whereas, lines which are very light or lightly drawn will appear to be part of the figure itself.

The lighting of any picture should, of course, always be considered. The greatest illumination is arrived at by keeping all details out of the lighted side, putting the shadows in also without detail. The more simply the forms are expressed, the more light will be shown.

It is generally a good idea to put details only in a picture in the half tones. This will give broadness and strength.

Distances are expressed in black and white in numbers of different ways. If the picture contains only black and white, distances can be shown by making the objects diminish in size as they recede. If the half tones are to be used, you will find that the lights must be darker and the darks must be lighter as they diminish and disappear into the distance. Likewise, the darks are darker and the lights are lighter as they advance to the foreground.

In closing this chapter, I wish to call to the attention of the artist that the entire work must be in the same atmosphere. To see an object very much stronger than its entire surroundings does not satisfy; it must be a part of the whole. If the picture is expressed in dark tones, many of the objects can be dark, but if expressed in light tones, it is not good to make a staccato note of one extreme dark.

Charcoal will be found to be the easiest medium with which to try out black-and-white mass composition.

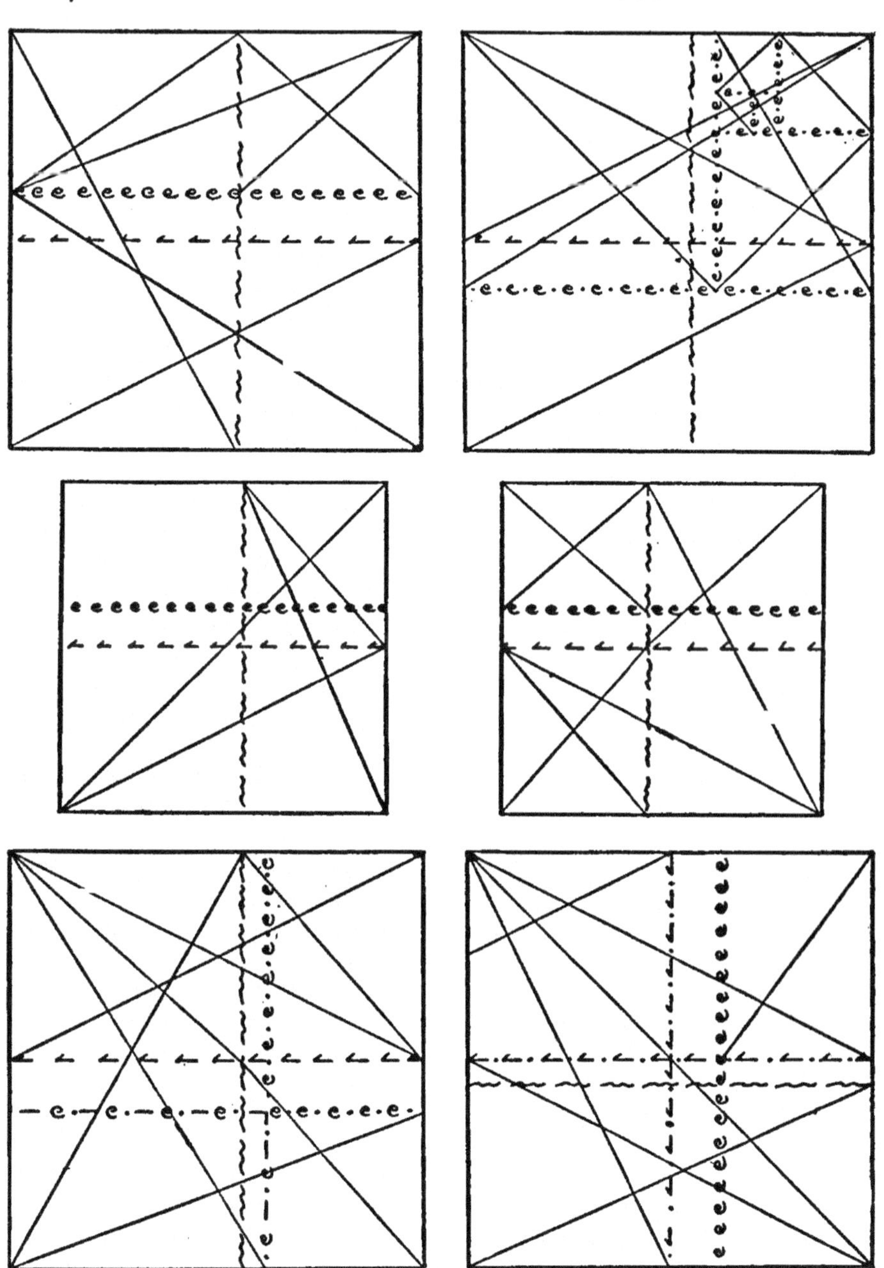

LAYOUTS OF BEN DAY ILLUSTRATIONS

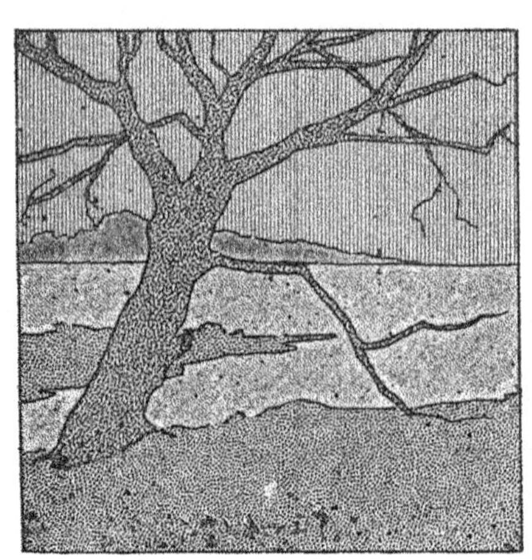

BEN DAY TEXTURES

CHAPTER FIFTEEN: COMPOSITION OF COLOUR

ATURE does not paint in black and white. All the world is full of colour. Even the blackest material will reflect some colour, and when we make a drawing in black and white, we have only said part of the truth; we have only given a conventional idea of our whole conception. We can also express an idea in black and white and one other colour. This also is only a conventional adaptation to suggest to the beholder a little more of our conception.

It was mentioned in the previous chapter that, besides painting objectively, which is painting nature as it really is, or as near as our pigments will permit us, there is another form of painting which, perhaps, is more æsthetic, which is getting the spirit of the thing, in colours which are only possible in our imagination, in a subjective style. It is the form of art which the Japanese, Chinese, Egyptians, and other peoples adapted and developed thousands of years ago. To-day our occidental minds are striving to work out this phase of art. Perhaps it is the kernel of all the modern schools.

In this subject form of painting, our imaginations carry us into the Elysian Fields; we can make ourselves as one of the gods, we can create; we can become a builder of things that do not already exist. Besides creating forms in this theme of art, we can originate in colours that are indeed strange, we can arrange colours contrary to all our previous coceptions of how nature is painted. We can make a sky: green, purple, red, blue, yellow, any colour, in fact, that we desire. We can make our trees any colour in the spectrum. We must be careful in doing this that our creations are based on nature's laws. If the artist will create original conceptions on the Dynamic Symmetry lines (as I have explained in the foregoing chapters) and will base his colour combinations also on nature's laws, in harmonies, in contrast, in split complementaries, etc., the sensation will be agreeable to the beholder.

In "The Art of Colour" and "The Study of Colour," I have gone deeper into this subject. Colour combinations are nothing more than colour compositions. Those who have read these books will understand what I mean when I say that a picture can be in two colours and two complementaries, or three colours and three complementaries, or a harmony of three, four, or five, or a split complementary of one against two, one against three, two against three, and one against six. These combinations and all the thousands that I have mentioned in the two books will be found to be pleasing. Of course, these colour combinations often are conventional arrangements, but this is the new art which many are striving to create.

If we are painting objectively, colour arrangements must conform more or less to the object we are painting from, but in subject painting or object painting, we must always be careful that the main points of interest and the other points of interest attract our eye in SEQUENCE, for the eye does not like to be distracted by two or more principal points of interest in one picture, except in certain large decorations where it is impossible to take in the whole picture at one glance.

Certain colours will always attract our eye more than others, especially as our pigments are not all as brilliant as nature's own spectrum. The red and orange, perhaps, are the two colours that first attract our eye. The yellow and scarlet would rank next. The crimson, yellow-green, and green next; the purple and blue-green next; the blue next; and the violet and blue-violet last. These are called sometimes advancing and receding colours. All colours can be made to advance or recede by surrounding them with other colours that are complementary, of more or less brilliancy.

Aërial perspective has also to do with colour composition. We must always remember the atmosphere changes all colours; and that the object, as it recedes, will partake somewhat of the colour of the atmosphere; the lighted side does not do so as quickly as the shadow. Then, again, it makes a difference what the colour of the object is, for a red object would have a purple shadow: this shadow would disappear more quickly into the distance than would the red, whereas the blue-violet object would vanish into the atmosphere colour more quickly than would the shadowy side. For a fuller explanation of this phenomenon I would refer you to Chapter Seven of "The Art of Colour."

How much of each colour to use in a picture depends on which colour we are going to paint the principal point of interest and the points in SEQUENCE. If we are using harmonies, it would be pleasing to use an even distribution of all colours, but if we are making a composition in contrasting colours, we can either paint the large masses in harmony and the complementary colours in the smaller masses, or we can put a colour with its complementary in juxtaposition. This holds good also with split complementaries where the larger number of colours can be used in harmony in the large masses and the smaller number of split complementaries in the smaller masses, or each set of complementaries with their mutual complementary.

If a colour is neutralized or gray and we wish to have an even distribution of weight of colour, we could use larger masses of the gray or neutralized colour and smaller masses of the brilliant tone.

If an object is brilliantly coloured in a large mass, all the surrounding objects will be influenced by this colour, and all other colours around it will partake of a colour *toward* the complementary in the same way that a shadow also partakes of a colour toward its complementary, clockwise or counter-clockwise on the spectrum circle.

CHAPTER SIXTEEN: A FEW MATHEMATICS OF DYNAMIC SYMMETRY

HIS chapter is written only for those who wish to study Dynamic Symmetry from a geometrical point of view and prove the correctness of the theory. It is not essential to the artist.

As I explained in the Foreword and Introduction of this book, I purposely left out any mathematical or geometrical reference in explaining Dynamic Symmetry. The task has been, perhaps, a little more difficult, on account of my determination not to use any signs, letters, or anything that might be misconstrued by the casual reader, by which he might be led to believe that Dynamic Symmetry was purely mathematics.

But in this chapter I am going to give some very simple data to those who wish to delve deeper into the "whys and wherefores." Of course, it will be impossible to go as deeply into this subject as Jay Hambidge has done in his book called "The Greek Vase" or in "Elements of Dynamic Symmetry," but perhaps those who have studied the preceding chapters will be better able to take up the task which apparently seems so difficult at the outset. After the first understanding of Dynamic Symmetry, the books of Jay Hambidge and Samuel Colman are wonderful explanations of the Greek form of composition, and I trust that some of the readers of this book will study and verify the compositional layouts illustrated in "The Greek Vase" and in "Elements of Dynamic Symmetry."

In the first chapter, the different roots were explained and how to form them, and how, by means of the diagonal, to make the smaller forms in summation. I took the liberty of naming them "SEQUENCE OF FORM." The basis of the area of each root is measured in what is known as square root. Of course, all those who have studied geometry and algebra know that the square root of 1 is 1. This is called Root One. A quantity which, taken twice as a factor, will produce the given quantity. Thus the square root of 25 is 5, because $5 \times 5 = 25$; so also $\frac{2}{3}$ is the square root of $\frac{4}{9}$, since $\frac{2}{3} \times \frac{2}{3} = \frac{4}{9}$; x^2 is the square root of x^4 since $x^2 \times x^2 = x^4$; $A + X$ is the square root of $a^2 + 2ax + x^2$, and so on. When the square root of a number can be expressed in exact parts of one, that number is a perfect square, and the indicated square root is said to be commensurable. All other indicated square roots are incommensurable. In other words, the square root is one of two equal factors of a given number. Thus 2 is the square root of 4, x of x^2. The following illustrates the method of finding the square root of 576, which is 24:

$$
\begin{array}{rrr}
 & \overline{\sqrt{576}} & (20 \\
 & 400 & 4 \\
 & \overline{} & \overline{} \\
2 \times 20 = 40) & 176 & 24 \\
(40 + 4) \times 4 = & 176 & \\
 & \overline{} &
\end{array}
$$

Quantities which when multiplied together produce unity are called, in mathematics, Reciprocals. Thus, the reciprocal of 1 is 1. The reciprocal of 2 is .50, and also, reversed, .50 is the reciprocal of 2. In other words, the reciprocal of a quantity is the quotient resulting from the division of unity by the quantity.

The square of Root Two is one side multiplied by unity to get the area of the rectangle, and we find that the reciprocal is 1.4142^+, which is an indeterminate fraction.

Taking the side of the rectangle as one (this would not necessarily mean one inch, or one mile, or one yard, but simply that you take the short side of the rectangle as the unit), you would find that the long side measured 1.4142^+ times the short side. The number 1.4142^+ is called the reciprocal, because, multiplied by itself, it would give you the Root Two. We must always remember that Dynamic Symmetry deals with areas and not with line. The Root Two rectangle contains two rectangles of the same shape in a smaller SEQUENCE. In other words, its reciprocal is equal to half the whole. (See Fig. 151.) Likewise, the major form is the reciprocal of the smaller form in SEQUENCE. In the same way, one half is the reciprocal of 2 and 2 is the reciprocal of the half.

Another way to show the relationship of form would be to draw a Root Two rectangle and alongside of this Root Two draw a square. Now, if we draw a small square which measures the width of the Root Two on the side, we will find that this smaller square will be exactly one half in area to the large square, and we see that, while Root Two is incommensurable in line, it is commensurable in area. (See Fig. 152.) So we see that the lesser square is the reciprocal of the greater square, and likewise, the greater square is the reciprocal of the lesser square. Thus, though the ends and

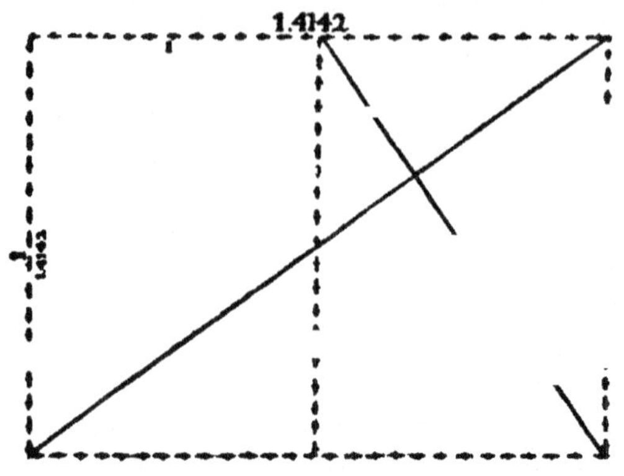

FIG. 151. THE RECIPROCAL OF ROOT TWO

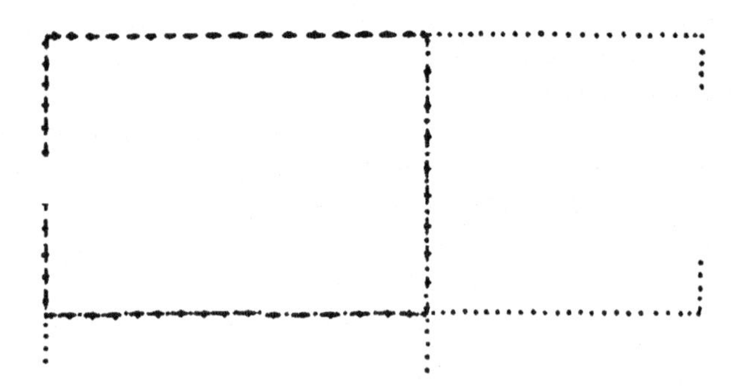

FIG. 152. RELATION OF MASS

the side are incommensurable in line, they are commensurable in area. The ends and side of this rectangle is 1 or unity to the square root of 2, or 1 to 1.4142^{+} which is an indeterminate fraction.

Therefore, we find that all the roots can be designated in the same manner as we have just designated Root Two. Before giving you the reciprocal of all the roots, I would like to show you the reciprocal of Root Two which is 1.4142^{+} multiplied by itself as follows:

$$1.4142^{+} \times 1.4142^{+} =$$

```
   1.4142+
   1.4142+
   -------
     28284
    56568
   14142
  56568
 14142
 ---------
 1.99996164
```

The small discrepancy is, of course, accounted for by the end of the fraction being dropped.

The reciprocal of all roots is given below. In other words, if any one of these numbers is multiplied by itself, it will give you the root, i. e. $1.4142^{+} \times 1.4142^{+} = 2$ or $2 \times 2 = 4$.

The reciprocal of Root One is	1.000
The reciprocal of Root Two is	1.414^{+}
The reciprocal of the Whirling Square Root is	1.618^{+}
The reciprocal of Root Three is	1.732^{+}
The reciprocal of Root Four is	2.000
The reciprocal of Root Five is	2.236^{+}
The reciprocal of Root Six is	2.449^{+}
The reciprocal of Root Seven is	2.645^{+}
The reciprocal of Root Eight is	2.828^{+}
The reciprocal of Root Nine is	3.000
The reciprocal of Root Ten is	3.162^{+}

The reciprocal of Root Eleven is . . . . 3.316+
The reciprocal of Root Twelve is . . . . 3.464+
The reciprocal of Root Thirteen is . . . . 3.605+
The reciprocal of Root Fourteen is . . . . 3.741+
The reciprocal of Root Fifteen is . . . . 3.872+
The reciprocal of Root Sixteen is 4.000
The reciprocal of Root Seventeen is . . . . 4.123+
The reciprocal of Root Eighteen is . . . . 4.242+
The reciprocal of Root Nineteen is . . . . 4.358+
The reciprocal of Root Twenty is . . . . 4.472+
The reciprocal of Root Twenty-one is . . . . 4.582+
The reciprocal of Root Twenty-two is . . . . 4.690+
The reciprocal of Root Twenty-three is . . . . 4.795+
The reciprocal of Root Twenty-four is . . . . 4.898+
The reciprocal of Root Twenty five is . . . . 5.000+

It must always be kept in mind that these numbers do not mean inches, centimetres, or squares, but mean the proportion which is always the same; whether in the major shape or in the minor shapes or forms in SEQUENCE, the geometrical relation will be constant. As we have explained, a rectangle whose side is expressed by the unit one and whose other side is expressed by 1.4142 or the square root of two, in which a diagonal or hypotenuse is drawn, both angles thus formed are equal; and if a crossing line is drawn through this hypotenuse so as to form four right angles, and a parallel line drawn where this line meets the side of the rectangle, the form so constructed will be also a Root Two.

This diagonal or hypotenuse or oblique line and the crossing line, together with the parallel, is a very important element in Dynamic Symmetry and can be reproduced in all roots. It will be noticed that, in all roots, this major rectangle and minor rectangle will be exactly in the proportion of the square root, i. e.: In Root Two the smaller square will be exactly one half. In the Root Three it will be one third. In the Root Four it will be one fourth, and in the Root Five, one fifth, as is illustrated on page 135.

The ratio of 1.618+ is used with unity to make a rectangle, which is divided by a diagonal, crossing lines, and the parallel to make a Whirling Square which is based on nature's design and which will make the logarithmetic curve. This can be done in two ways within the root of the Whirling

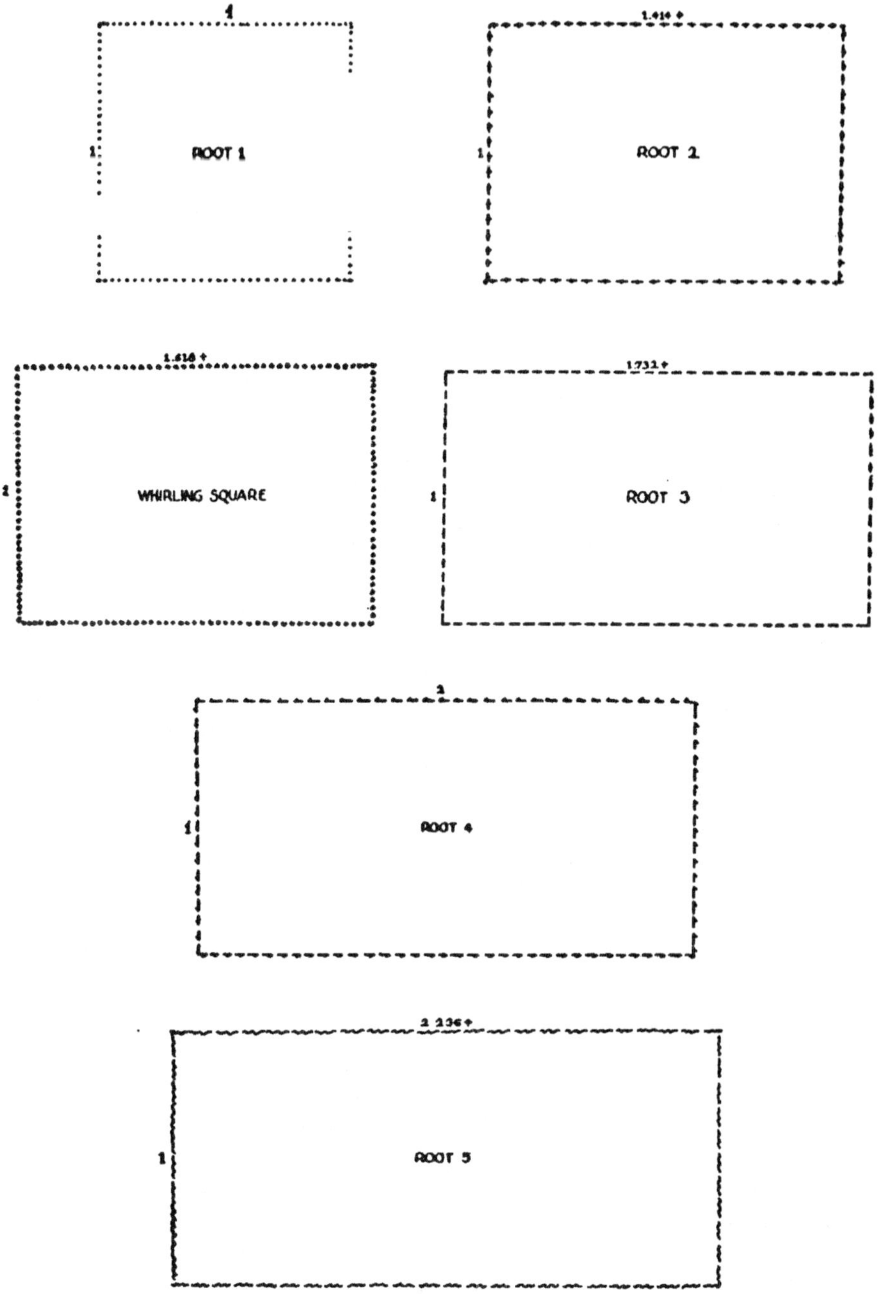

1

1 ROOT 1

1.414+

1 ROOT 2

1.618+

1 WHIRLING SQUARE

1.732+

1 ROOT 3

2

1 ROOT 4

2.236+

1 ROOT 5

RECIPROCALS OF ALL ROOTS

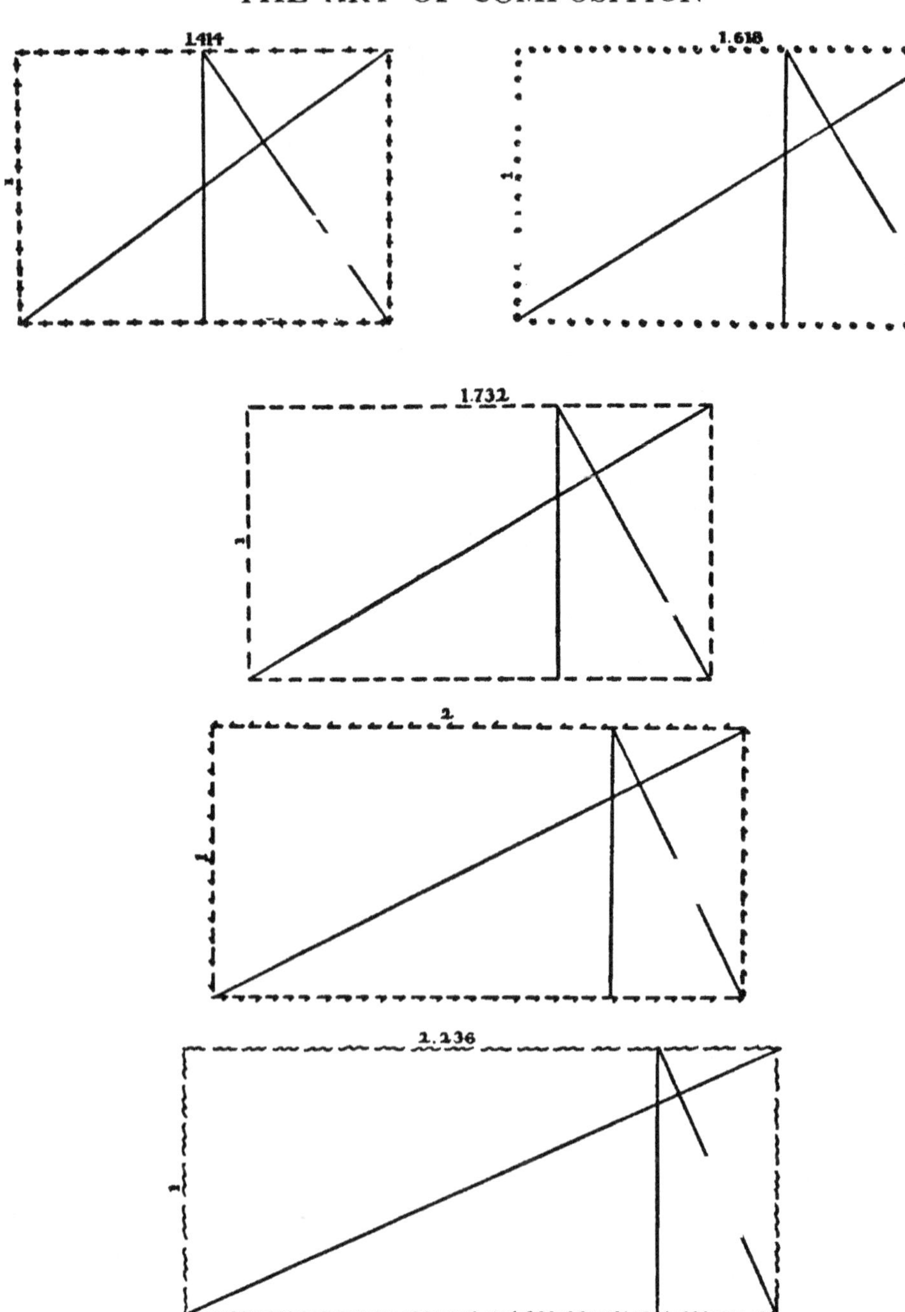

THE SQUARE ROOT OF ALL ROOT RECTANGLES

Square, and also outside the rectangle, whose reciprocal is 1.618+. If you will draw a square whose side measures 4.50 centimetres, laying off half the square and finding the diagonal of this half, adding this to half the square to make the Whirling Square Root, you will have a rectangle whose side will measure 7.236 centimetres. If we draw the diagonal and the crossing lines together with the parallel line, bringing down the forms in SEQUENCE, and draw the Whirling Square around the *outside* of the rectangle, as is shown in Fig. 153, we shall have completed the logarithmetic spiral, or Whirling Square, outside the rectangle.

If we now draw another square which measures 5 centimetres and make the Whirling Square in the same way as we did in the preceding paragraph, we shall have a rectangle whose side measures 8.20 centimetres plus or minus.

Now, if we draw again the diagonals and the crossing lines and draw the Whirling Square inside the rectangle connecting up the hypotenuse of each square, we shall have a Whirling Square which will measure identically with the Whirling Square in the preceding paragraph, but turned obliquely, as is shown in Fig. 154. Both of these rectangles have a ratio of 1.618+.

It is much more important to the artist to be able to do this inside of the Major Form, although, in "The Greek Vase," Jay Hambidge did not show this, to my knowledge.

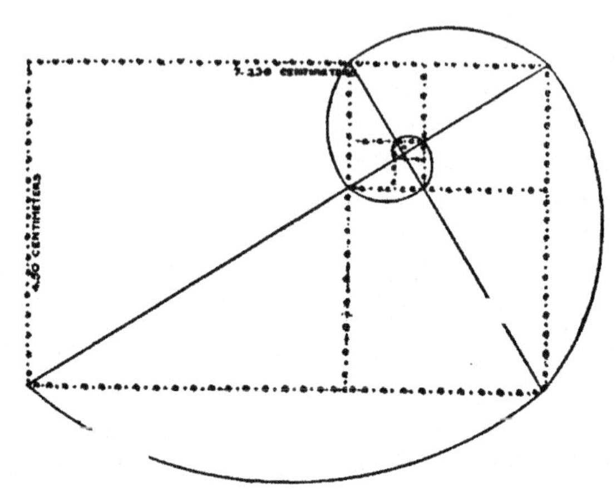

FIG. 153. WHIRLING SQUARE ROOT OUTSIDE OF THE RECTANGLE

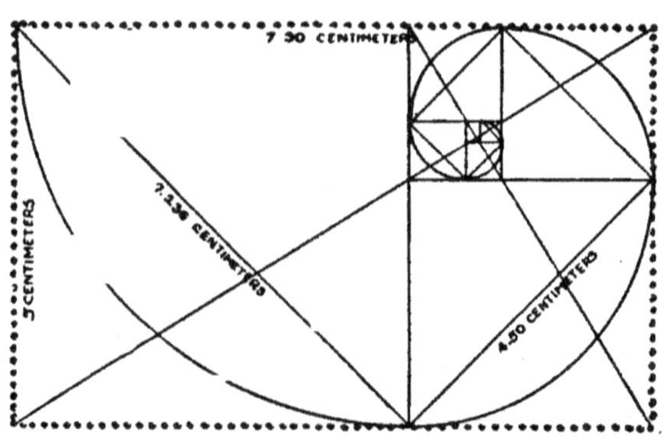

FIG. 154. WHIRLING SQUARE ROOT INSIDE OF THE RECTANGLE

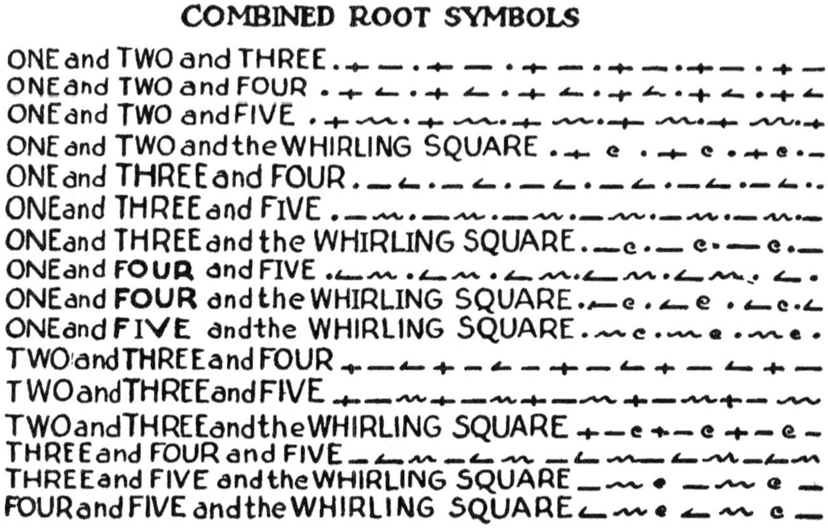

COMBINED ROOT SYMBOLS

ONE and TWO and THREE
ONE and TWO and FOUR
ONE and TWO and FIVE
ONE and TWO and the WHIRLING SQUARE
ONE and THREE and FOUR
ONE and THREE and FIVE
ONE and THREE and the WHIRLING SQUARE
ONE and FOUR and FIVE
ONE and FOUR and the WHIRLING SQUARE
ONE and FIVE and the WHIRLING SQUARE
TWO and THREE and FOUR
TWO and THREE and FIVE
TWO and THREE and the WHIRLING SQUARE
THREE and FOUR and FIVE
THREE and FIVE and the WHIRLING SQUARE
FOUR and FIVE and the WHIRLING SQUARE

FIG. 155

GLOSSARY

Balance: A picture so arranged that the objects balance each other either dynamically or statically. A composition which gives us a satisfaction of mechanical rest or continued movement.

Commensurable: Measurable by a common unit. Proportionate.

Crossing line: A line drawn from one corner the opposite from which the diagonal is drawn, through the diagonal line touching the outer side of a rectangle.

Diagonal: A line drawn from two opposite corners. An oblique line. The hypotenuse of a rectangle. A straight line showing two opposite vertices.

Dynamic Symmetry: A form of composition and proportion which was used by the Greeks and Egyptians.

Horizontal: On a level. In the direction of or parallel to the horizon. Flat. Plane.

Hypotenuse: The long side of a right-angled triangle.

Incommensurable: Not measurable by a common unit. Opposite of commensurable.

Logarithmetic: A curve which progresses in width from the centre or eye in a certain ratio. This ratio is based on nature's law of growth as found by the Royal Botanical Society. See illustration of sunflower, page 11.

Oblong: Having one principal axis longer than the other or others.

Parallel: Not meeting or intersecting, how far soever extended: said of straight lines or planes.

Parallel line: Any line, either perpendicular or horizontal which is parallel to the sides or ends of a rectangle.

Parallelogram: A quadrilateral whose opposite sides are parellel.

Perpendicular: Upright or vertical.

Points of interest: The place in a picture which first attracts our eye. The object which shows the important conception. It is found at the intersection of the crossing line and the diagonal of any rectangle in Dynamic Symmetry.

Second point of interest: The object which we next are conscious of after seeing the principal point of interest. It should not attract our eye before the principal point of interest does. It can be placed at the intersection of any crossing line in SEQUENCE to the principal point of interest.

Third point of interest: The object which is placed in SEQUENCE to the first and second point of interest.

Fourth point of interest: The object which leads us from the major points of interest.

Fifth point of interest: An object of not much importance which leads us in SEQUENCE from the major points of interest.

Quadrant arc: A part of a circle, the two ends resting in opposite corners to the diagonal.

Quotient: The result obtained by division; in arithmetic, a number indicating how many times one number or quantity is contained in another.

Rectangle: A parallelogram whose angles are right angles. An oblong or square.

Right angle: An angle formed by two straight lines which intersect each other perpendicularly.

Root One: A square 1 × 1. Its reciprocal is 1. The Root One can contain any one of the other roots. Dividing it in half will make two Root Fours. The other roots are found inside of a square by means of a quadrant arc or quarter circle and the diagonal.

Root Two: A rectangle which can be divided into two equal parts, both parts forming Root Two in SEQUENCE or lesser magnitude. Its reciprocal is 1.414+. The diagonal of Root One is the length of Root Two.

Root Three: A rectangle which can be divided into three equal parts, each one of the three parts forming a Root Three in SEQUENCE or form of lesser magnitude, in the same proportion. Its reciprocal is 1.732+. The diagonal of Root Two is the length of Root Three.

Root Four: A rectangle which can be divided into four parts each one of the four parts forming a Root Four in SEQUENCE or form of lesser magnitude, in the same proportion. Its reciprocal is 2. The diagonal of Root Three is the length of Root Four.

Root Five: A rectangle which can be divided into five equal parts each one of the five parts forming a Root Five in SEQUENCE or form of lesser magnitude, in the same proportion. Its reciprocal is 2.236+. The diagonal of Root Four is the length of Root Five.

Roots Outside of a Square: These are found by measuring the diagonal, taking this as the length of the Root Two.

Rule of Three: The product of the extremes is equal to the product of the means.

Sequence: The process of following in numbers each related to the other. A number of things related to each other considered collectively. A series.

Square: A figure having four equal sides and four right angles. A rectangle whose sides are equal.

Square Root: A quantity which, being taken twice as a factor, will produce the given quantity. Thus, the square root of 25 is 5, because $5 \times 5 = 25$. When the square root of a number can be expressed in exact parts of 1, that number is a perfect square, and the indicated square root is said to be commensurable. All other indicated square roots are incommensurable. One of two equal factors of a given number. Thus 2 is the square root of 4, x of x^2.

Star layout: Inside a rectangle lines drawn from opposite corners or hypotenuse and crossing lines drawn at right angles to the hypotenuse.

Summation: A form of numbering which is adding the preceding number to the following number such as 1, 2, 3, 5, 8, 13, 21, or 1.4142, 2.4142, 3.8284, 6.2426.

Symmetry: A due proportion of several parts of a body to each other, or the union and conformity of the members of a work to the whole. Symmetry arises from the proportion, which the Greeks called analogy, which is the relation of conformity of all parts of certain measure.

Whirling Square Root: This root is found by taking half of a square, drawing the diagonal across this half, and adding this to half the width of the square. A Root Five contains two Whirling Square Roots and a square. The Whirling Square is the form of the logarithmetic curve which is based on Nature's growth and leaf distribution.

The Study of Colour

With Lessons and Exercises

2nd Edition in Two Volumes

by

Michel Jacobs

For the Artist *For the Student* *For the Craftsman*

The lessons in this book are the development of years of experience in the Metropolitan Art School, the result of careful selection, graded in order of difficulty, and covering the subject thoroughly and carefully. It is a theoretical textbook on colour with practical lessons and exercises, and covers the course taught today in the Metropolitan Art School of New York.

Those who are familiar with Michel Jacobs's first book, "The Art of Colour," cannot afford to be without this new work. "It is no longer a theory, it is a proven fact."

The price has been made moderate so as to be within the reach of all.

Vol. I. Lessons, Bound Separately....$2.50 For Instructors
Vol. II. Charts, Bound Separately..... 1.00 For Classes
Vol. I and II. Bound in One Volume... 3.00

If the art student would execute the lessons which the book includes, he is bound to see colour and to learn how to use it in his work.

Five hundred thousand colour combinations.
Sixteen coloured illustrations.
Sixty-four black and white illustrations.
All charts for colouring.

Published by

Doubleday, Page & Company

Garden City New York

CPSIA information can be obtained
at www.ICGtesting.com
Printed in the USA
LVOW13s0958010917
547234LV00019B/121/P

9 780282 561055